Flinn Scientific ChemTopic™ Labs

Solids and Liquids

Senior Editor

Irene Cesa
Flinn Scientific, Inc.
Batavia, IL

Curriculum Advisory Board

Bob Becker
Kirkwood High School
Kirkwood, MO

Kathleen J. Dombrink
McCluer North High School
Florissant, MO

Robert Lewis
Downers Grove North High School
Downers Grove, IL

John G. Little
St. Mary's High School
Stockton, CA

Lee Marek
University of Illinois–Chicago
Chicago, IL

John Mauch
Braintree High School
Braintree, MA

Dave Tanis
Grand Valley State University
Allendale, MI

FLINN SCIENTIFIC INC.
"Your Safer Source for Science Supplies"
P.O. Box 219 • Batavia, IL 60510
1-800-452-1261 • www.flinnsci.com

ISBN 1-877991-79-1

Copyright © 2005 Flinn Scientific, Inc.

All rights reserved. No part of this book may be reproduced or transmitted in any form or by any means, electronic or mechanical, including, but not limited to photocopy, recording, or any information storage and retrieval system, without permission in writing from Flinn Scientific, Inc.
No part of this book may be included on any Web site.

Reproduction permission is granted only to the science teacher who has purchased this volume of Flinn ChemTopic™ Labs, Solids and Liquids, Catalog No. AP6660 from Flinn Scientific, Inc. Science teachers may make copies of the reproducible student pages for use only by their students.

Printed in the United States of America.

Table of Contents

	Page
Flinn ChemTopic™ Labs Series Preface	i
About the Curriculum Advisory Board	ii
Solids and Liquids Preface	iii
Format and Features	iv–v
Experiment Summaries and Concepts	vi–vii
Additional Activities	viii

Experiments

It's Just a Phase	1
Properties of Liquids	13
Vapor Pressure of Water	25
How Cool Is That?	35
Teaching with Toys	49

Demonstrations

Hot Wax	53
"Tennis Ball" Distillation	55
Surface Tension Jar	57
Freezing by Boiling	59
Wet Dry Ice	61
Four-Square Diffusion	63

Supplementary Information

Safety and Disposal Guidelines	66
National Science Education Standards	68
Master Materials Guide	70

Flinn ChemTopic™ Labs Series Preface
Lab Manuals Organized Around Key Content Areas in Chemistry

In conversations with chemistry teachers across the country, we have heard a common concern. Teachers are frustrated with their current lab manuals, with experiments that are poorly designed and don't teach core concepts, with procedures that are rigid and inflexible and don't work. Teachers want greater flexibility in their choice of lab activities. As we further listened to experienced master teachers who regularly lead workshops and training seminars, another theme emerged. Master teachers mostly rely on collections of experiments and demonstrations they have put together themselves over the years. Some activities have been passed on like cherished family recipe cards from one teacher to another. Others have been adapted from one format to another to take advantage of new trends in microscale equipment and procedures, technology innovations, and discovery-based learning theory. In all cases the experiments and demonstrations have been fine-tuned based on real classroom experience.

Flinn Scientific has developed a series of lab manuals based on these "cherished recipe cards" of master teachers with proven excellence in both teaching students and training teachers. Created under the direction of an Advisory Board of award-winning chemistry teachers, each lab manual in the Flinn ChemTopic™ Labs series contains 4–6 student-tested experiments that focus on essential concepts and applications in a single content area. Each lab manual also contains 4–6 demonstrations that can be used to illustrate a chemical property, reaction, or relationship and will capture your students' attention. The experiments and demonstrations in the Flinn ChemTopic™ Labs series are enjoyable, highly focused, and will give students a real sense of accomplishment.

Laboratory experiments allow students to experience chemistry by doing chemistry. Experiments have been selected to provide students with a crystal-clear understanding of chemistry concepts and encourage students to think about these concepts critically and analytically. Well-written procedures are guaranteed to work. Reproducible data tables teach students how to organize their data so it is easily analyzed. Comprehensive teacher notes include a master materials list, solution preparation guide, complete sample data, and answers to all questions. Detailed lab hints and teaching tips show you how to conduct the experiment in your lab setting and how to identify student errors and misconceptions before students are led astray.

Chemical demonstrations provide another teaching tool for seeing chemistry in action. Because they are both visual and interactive, demonstrations allow teachers to take students on a journey of observation and understanding. Demonstrations provide additional resources to develop central themes and to magnify the power of observation in the classroom. Demonstrations using discrepant events challenge student misconceptions that must be broken down before new concepts can be learned. Use demonstrations to introduce new ideas, illustrate abstract concepts that cannot be covered in lab experiments, and provide a spark of excitement that will capture student interest and attention.

Safety, flexibility, and choice

Safety always comes first. Depend on Flinn Scientific to give you upfront advice and guidance on all safety and disposal issues. Each activity begins with a description of the hazards involved and the necessary safety precautions to avoid exposure to these hazards. Additional safety, handling, and disposal information is also contained in the teacher notes.

The selection of experiments and demonstrations in each Flinn ChemTopic™ Labs manual gives you the flexibility to choose activities that match the concepts your students need to learn. No single teacher will do all of the experiments and demonstrations with a single class. Some experiments and demonstrations may be more helpful with a beginning-level class, while others may be more suitable with an honors class. All of the experiments and demonstrations have been keyed to national content standards in science education.

Chemistry is an experimental science!

Whether they are practicing key measurement skills or searching for trends in the chemical properties of substances, all students will benefit from the opportunity to discover chemistry by doing chemistry. No matter what chemistry textbook you use in the classroom, Flinn ChemTopic™ Labs will help you give your students the necessary knowledge, skills, attitudes, and values to be successful in chemistry.

About the Curriculum Advisory Board

Flinn Scientific is honored to work with an outstanding group of dedicated chemistry teachers. The members of the Flinn ChemTopic Labs Advisory Board have generously contributed their proven experiments and demonstrations to create these topic lab manuals. The wisdom, experience, creativity, and insight reflected in their lab activities guarantee that students who perform them will be more successful in learning chemistry. On behalf of all chemistry teachers, we thank the Advisory Board members for their service and dedication to chemistry education.

Bob Becker teaches chemistry and AP chemistry at Kirkwood High School in Kirkwood, MO. Bob received his B.A. from Yale University and M.Ed. from Washington University and has 20 years of teaching experience. A well-known demonstrator, Bob has conducted more than 100 demonstration workshops across the U.S. and Canada and was a Team Leader for the Flinn Foundation Summer Workshop Program. His creative and unusual demonstrations have been published in the *Journal of Chemical Education,* the *Science Teacher,* and *Chem13 News.* Bob is the author of two books of chemical demonstrations, *Twenty Demonstrations Guaranteed to Knock Your Socks Off, Volumes I and II,* published by Flinn Scientific. Bob has been awarded the James Bryant Conant Award in High School Teaching from the American Chemical Society, the Regional Catalyst Award from the Chemical Manufacturers Association, and the Tandy Technology Scholar Award.

Kathleen J. Dombrink teaches chemistry and advanced-credit college chemistry at McCluer North High School in Florissant, MO. Kathleen received her B.A. in Chemistry from Holy Names College and M.S. in Chemistry from St. Louis University and has 35 years of teaching experience. Recognized for her strong support of professional development, Kathleen has been selected to participate in the Fulbright Memorial Fund Teacher Program in Japan and NEWMAST and Dow/NSTA Workshops. She served as co-editor of the inaugural issues of *Chem Matters* and was a Woodrow Wilson National Fellowship Foundation Chemistry Team Member for 11 years. Kathleen is currently a Team Leader for the Flinn Foundation Summer Workshop Program. Kathleen has received the Presidential Award, the Midwest Regional Teaching Award from the American Chemical Society, the Tandy Technology Scholar Award, and a Regional Catalyst Award from the Chemical Manufacturers Association.

Robert Lewis retired from teaching chemistry at Downers Grove North High School in Downers Grove, IL, and is currently a Secondary Coordinator for the GATE program in Chicago. Robert received his B.A. from North Central College and M.A. from University of the South and has 30 years of teaching experience. He was a founding member of Weird Science, a group of chemistry teachers that traveled throughout the country to stimulate teacher enthusiasm for using demonstrations to teach science. Robert served as a Team Leader for both the Woodrow Wilson National Fellowship Foundation and the Flinn Foundation Summer Workshop Program. Robert has received the Presidential Award, the James Bryant Conant Award in High School Teaching from the American Chemical Society, the Tandy Technology Scholar Award, a Regional Catalyst Award from the Chemical Manufacturers Association, and a Golden Apple Award from the State of Illinois.

John G. Little teaches chemistry and AP chemistry at St. Mary's High School in Stockton, CA. John received his B.S. and M.S. in Chemistry from University of the Pacific and has 39 years of teaching experience. Highly respected for his well-designed labs, John is the author of two lab manuals, *Chemistry Microscale Laboratory Manual* (D. C. Heath), and *Microscale Experiments for General Chemistry* (with Kenneth Williamson, Houghton Mifflin). He is also a contributing author to *Science Explorer* (Prentice Hall) and *World of Chemistry* (McDougal Littell). John served as a Chemistry Team Leader for both the Woodrow Wilson National Fellowship Foundation and the Flinn Foundation Summer Workshop Program. He has been recognized for his dedicated teaching with the Tandy Technology Scholar Award and the Regional Catalyst Award from the Chemical Manufacturers Association.

Lee Marek retired from teaching chemistry at Naperville North High School in Naperville, IL and currently teaches at the University of Illinois–Chicago. Lee received his B.S. in Chemical Engineering from the University of Illinois and M.S. degrees in Physics and Chemistry from Roosevelt University. He has more than 30 years of teaching experience and is a Team Leader for the Flinn Foundation Summer Workshop Program. His students have won national recognition in the International Chemistry Olympiad, the Westinghouse Science Talent Search, and the Internet Science and Technology Fair. Lee was a founding member of Weird Science and has presented more than 500 demonstration and teaching workshops for more than 300,000 students and teachers across the country. Lee has performed science demonstrations on the *David Letterman Show* 20 times. Lee has received the Presidential Award, the James Bryant Conant Award in High School Teaching and the Helen M. Free Award for Public Outreach from the American Chemical Society, the National Catalyst Award from the Chemical Manufacturers Association, and the Tandy Technology Scholar Award.

John Mauch teaches chemistry and AP chemistry at Braintree High School in Braintree, MA. John received his B.A. in Chemistry from Whitworth College and M.A. in Curriculum and Education from Washington State University and has more than 25 years of teaching experience. John is an expert in microscale chemistry and is the author of two lab manuals, *Chemistry in Microscale, Volumes I and II* (Kendall/Hunt). He is also a dynamic and prolific demonstrator and workshop leader. John has presented the Flinn Scientific Chem Demo Extravaganza show at NSTA conventions for eight years and has conducted more than 100 workshops across the country. John was a Chemistry Team Member for the Woodrow Wilson National Fellowship Foundation program and is currently a Board Member for the Flinn Foundation Summer Workshop Program. John has received the Massachusetts Chemistry Teacher of the Year Award from the New England Institute of Chemists.

Dave Tanis is Associate Professor of Chemistry at Grand Valley State University in Allendale, MI. Dave received his B.S. in Physics and Mathematics from Calvin College and M.S. in Chemistry from Case Western Reserve University. He taught high school chemistry for 26 years before joining the staff at Grand Valley State University to direct a coalition for improving precollege math and science education. Dave later joined the faculty at Grand Valley State University and currently teaches courses for pre-service teachers. The author of two laboratory manuals, Dave acknowledges the influence of early encounters with Hubert Alyea, Marge Gardner, Henry Heikkinen, and Bassam Shakhashiri in stimulating his long-standing interest in chemical demonstrations and experiments. Continuing this tradition of mentorship, Dave has led more than 40 one-week institutes for chemistry teachers and served as a Team Member for the Woodrow Wilson National Fellowship Foundation for 13 years. He is currently a Board Member for the Flinn Foundation Summer Workshop Program. Dave received the College Science Teacher of the Year Award from the Michigan Science Teachers Association.

Preface
Solids and Liquids

You don't have to know any chemistry to appreciate the beautiful crystalline structures of minerals and gems or to enjoy the cool relief of an ocean breeze. However, the properties of solids and liquids provide a mirror for us to "see inside" the world of atoms and molecules—to understand the motion of molecules and to compare the interactions among different types of molecules. The properties of solids and liquids offer persuasive and convincing evidence for the kinetic-molecular theory, the most important model for explaining the physical properties of matter. The purpose of *Solids and Liquids*, Volume 11 in the *Flinn ChemTopic™ Labs* series, is to provide high school chemistry teachers with laboratory activities that will help students explain and predict the properties of solids and liquids. Five experiments and six demonstrations allow students to investigate phase transitions, identify the accompanying energy changes, and recognize the underlying influence of attractive forces between molecules.

Energy and Phase Changes

Reversible phase changes involving solids and liquids include melting and freezing, evaporation and condensation, and sublimation. In the experiment "It's Just a Phase," students measure heating and cooling curves for lauric acid, a low-melting organic solid, identify the melting point, and interpret the energy changes that accompany solid–liquid phase changes. A quantitative adaptation of this experiment is used in the "Hot Wax" demonstration to determine the amount of heat required to melt a solid. The experiment "How Cool Is That?" looks at the "next" transition in the series, that from liquid to gas. Students measure the temperature changes that occur when different liquids evaporate and analyze the "cooling effect of evaporation" in terms of the existence of intermolecular forces. The ultimate cooling effect of evaporation is observed in the demonstration "Freezing by Boiling." Boiling a liquid at reduced pressure makes it cold enough to freeze—freezing and boiling occur simultaneously! All three phases, solid, liquid, and gas, are also observed simultaneously in the demonstration "Wet Dry Ice," an amazingly simple way to show the triple point of carbon dioxide.

Intermolecular Forces

The properties of solids and liquids depend on the nature and strength of intermolecular attractive forces between molecules. Nowhere is this more evident than in the structure and properties of water. Hydrogen bonds, the strongest type of intermolecular forces, give water its unique properties. In "Properties of Liquids," students observe the "forceful" effects of surface tension in water and compare the capillary action and surface tension of water versus other liquids. The "Surface Tension Jar" demonstration provides an alternative way to illustrate this interesting phenomenon.

Kinetic-Molecular Theory

The kinetic-molecular theory may be summarized in one simple phrase—molecules in motion. See the "Tennis Ball Distillation" activity for a great way to explain solid–liquid phase changes in terms of this important theory. A tub-full of tennis balls is a perfect demonstration device for comparing the energy and motion of molecules in the solid, liquid, and gas phases. Visualizing molecules in motion also helps explain why different liquids have different vapor pressures at the same temperature and why the vapor pressure of a liquid always increases as the temperature increases. In the experiment "Vapor Pressure of Water," students determine the vapor pressure of water at different temperatures by measuring the volume of "wet air" as a function of temperature. Students can also learn about vapor pressure by examining the chemical principles at work in two popular toys, the drinking bird and the hand boiler. "Teaching with Toys" reminds us all of something we may have forgotten in growing up—that learning is indeed child's play!

Safety and Success!

Chemistry is an experimental science! Depend on Flinn Scientific to give you the information and confidence you need to work safely with your students and help them succeed. As your safer source for science supplies, Flinn Scientific promises you the most reliable safety information for every potential lab hazard. Whether you are looking for an updated classic or a creative simulation, our labs offer you safe solutions and practical alternatives. The selection of experiments and demonstrations in *Solids and Liquids* gives you the ability to design an effective lab curriculum that will work with your students and your resources in your classroom. Best of all, no matter which activities you choose, your students are assured of success. All of the activities in *Solids and Liquids* have been thoroughly tested and retested. You know they will work! Use the experiment summaries and concepts on the following pages to locate the concepts you want to teach and to choose experiments and demonstrations that will help you meet your goals.

Format and Features

Flinn ChemTopic™ Labs

All experiments and demonstrations in Flinn ChemTopic™ Labs are printed in a 10⅞" × 11" format with a wide 2" margin on the inside of each page. This reduces the printed area of each page to a standard 8½" × 11" format suitable for copying.

The wide margin assures you the entire printed area can be easily reproduced without damaging the binding. The margin also provides a convenient place for teachers to add their own notes.

Concepts — Use these bulleted lists along with state and local standards, lesson plans, and your textbook to identify activities that will allow you to accomplish specific learning goals and objectives.

Background — A balanced source of information for students to understand why they are doing an experiment, what they are doing, and the types of questions the activity is designed to answer. This section is not meant to be exhaustive or to replace the students' textbooks, but rather to identify the core concepts that should be covered before starting the lab.

Experiment Overview — Clearly defines the purpose of each experiment and how students will achieve this goal. Performing an experiment without a purpose is like getting travel directions without knowing your destination. It doesn't work, especially if you run into a roadblock and need to take a detour!

Pre-Lab Questions — Making sure that students are prepared for lab is the single most important element of lab safety. Pre-lab questions introduce new ideas or concepts, review key calculations, and reinforce safety recommendations. The pre-lab questions may be assigned as homework in preparation for lab or they may be used as the basis of a cooperative class activity before lab.

Materials — Lists chemical names, formulas, and amounts for all reagents—along with specific glassware and equipment—needed to perform the experiment as written. The material dispensing area is a main source of student delay, congestion, and accidents. Three dispensing stations per room are optimum for a class of 24 students working in pairs. To safely substitute different items for any of the recommended materials, refer to the *Lab Hints* section in each experiment or demonstration.

Safety Precautions — Instruct and warn students of the hazards associated with the materials or procedure and give specific recommendations and precautions to protect students from these hazards. Please review this section with students before beginning each experiment.

Procedure — This section contains a stepwise, easy-to-follow procedure, where each step generally refers to one action item. Contains reminders about safety and recording data where appropriate. For inquiry-based experiments the procedure may restate the experiment objective and give general guidelines for accomplishing this goal.

Data Tables — Data tables are included for each experiment and are referred to in the procedure. These are provided for convenience and to teach students the importance of keeping their data organized in order to analyze it. To encourage more student involvement, many teachers prefer to have students prepare their own data tables. This is an excellent pre-lab preparation activity—it ensures that students have read the procedure and are prepared for lab.

Post-Lab Questions or Data Analysis — This section takes students step-by-step through what they did, what they observed, and what it means. Meaningful questions encourage analysis and promote critical thinking skills. Where students need to perform calculations or graph data to analyze the results, these steps are also laid out sequentially.

Format and Features
Teacher's Notes

Master Materials List — Lists the chemicals, glassware, and equipment needed to perform the experiment. All amounts have been calculated for a class of 30 students working in pairs. For smaller or larger class sizes or different working group sizes, please adjust the amounts proportionately.

Preparation of Solutions — Calculations and procedures are given for preparing all solutions, based on a class size of 30 students working in pairs. With the exception of particularly hazardous materials, the solution amounts generally include 10% extra to account for spillage and waste. Solution volumes may be rounded to convenient glassware sizes (100-mL, 250-mL, 500-mL, etc.).

Safety Precautions — Repeats the safety precautions given to the students and includes more detailed information relating to safety and handling of chemicals and glassware. Refers to Material Safety Data Sheets that should be available for all chemicals used in the laboratory.

Disposal — Refers to the current *Flinn Scientific Catalog/Reference Manual* for general guidelines and specific procedures governing the disposal of laboratory waste. Because we recommend that teachers review local regulations before beginning any disposal procedure, the information given in this section is for general reference purposes only. However, if a disposal step is included as part of the experimental procedure itself, then the specific solutions needed for disposal are described in this section.

Lab Hints — This section reveals common sources of student errors and misconceptions and where students are likely to need help. Identifies the recommended length of time needed to perform each experiment, suggests alternative chemicals and equipment that may be used, and reminds teachers about new techniques (filtration, pipeting, etc.) that should be reviewed prior to lab.

Teaching Tips — This section puts the experiment in perspective so that teachers can judge in more detail how and where a particular experiment will fit into their curriculum. Identifies the working assumptions about what students need to know in order to perform the experiment and answer the questions. Highlights historical background and applications-oriented information that may be of interest to students.

Sample Data — Complete, actual sample data obtained by performing the experiment exactly as written is included for each experiment. Student data will vary.

Answers to All Questions — Representative or typical answers to all questions. Includes sample calculations and graphs for all data analysis questions. Information of special interest to teachers only in this section is identified by the heading "Note to the teacher." Student answers will vary.

Look for these icons in the *Experiment Summaries and Concepts* section and in the *Teacher's Notes* of individual experiments to identify inquiry-, microscale-, and technology-based experiments, respectively.

Experiment Summaries and Concepts

Experiment

It's Just a Phase—Heating and Cooling Curves

It seems counter-intuitive—when freezing weather is predicted, orange growers spray the trees with water to protect the fruit from freezing. The water actually releases heat as it solidifies! The purpose of this experiment is to investigate the solid–liquid phase changes for lauric acid, an organic compound that is used to make soap. Students measure the heating and cooling curves for lauric acid and analyze the results to determine the melting point and the energy changes that take place when a liquid freezes or a solid melts.

Properties of Liquids—Surface Tension and Capillary Action

Water hardly seems special. It is, after all, a very common "chemical." Water is an unusual liquid, however, with unique physical properties, such as a very high surface tension. The purpose of this experiment is to observe the "forceful" effects of surface tension in water and to compare the capillary action and surface tension of water versus other liquids. Students investigate how the properties of liquids depend on the nature and strength of attractive forces between molecules.

Vapor Pressure of Water—Effect of Temperature

Dry air in buildings and homes causes many health problems during the winter months, because the amount of water vapor in the air depends on temperature. Why does warm air hold more water than cold air? The purpose of this experiment is to determine the vapor pressure of water at different temperatures. The vapor pressure will be measured by trapping a small amount of air in an inverted graduated cylinder and then measuring how the volume of gas changes as the temperature is reduced.

How Cool is That?—Evaporation of Liquids

The "cooling effect of evaporation" is nature's most important way of cooling not only our bodies but also the Earth. Water evaporating from the Earth's surface, for example, helps to moderate the temperature and climate around large bodies of water. How cool is evaporation? The purpose of this experiment is to measure the temperature changes that occur when different liquids evaporate and to compare their rates of evaporation. Liquids will be compared pair-wise and the results will be analyzed in terms of the strength of attractive forces between different types of molecules.

Teaching with Toys—Drinking Bird and Hand Boiler

Learning is child's play! Use this inquiry-based activity to let students discover the chemical principles at work (and in play!) for two popular toys, the "Drinking Bird" and the "Hand Boiler." The drinking bird demonstrates the cooling effect of evaporation and the effect of temperature on vapor pressure, while the hand boiler is actually a distillation apparatus in disguise!

Concepts

- Solids and liquids
- Phase changes
- Melting point
- Kinetic-molecular theory

- Properties of liquids
- Surface tension
- Capillary action
- Intermolecular forces

- Evaporation and condensation
- Vapor pressure
- Kinetic-molecular theory
- Ideal gas law and Dalton's law

- Evaporation
- Kinetic-molecular theory
- Polar vs. nonpolar compounds
- Hydrogen bonding

- Evaporation
- Vapor pressure
- Distillation

Experiment Summaries and Concepts

Demonstration

Hot Wax—Heat of Fusion of Paraffin

The temperature at which paraffin (candle wax) melts is only about 55 °C. When hot wax solidifies, however, it releases heat and causes severe skin burns. Use this calorimetry demonstration to measure the heat of fusion of paraffin and to compare the endothermic and exothermic nature of melting and freezing, respectively.

"Tennis Ball" Distillation—Kinetic-Molecular Theory in Action

Phase changes such as melting and evaporation are very "moving" events when molecules are the size of tennis balls! A tub-full of tennis balls is a perfect demonstration device for comparing the energy and motion of molecules in the solid, liquid, and gas phases. It's time to put the "kinetic" back into the kinetic-molecular theory!

Surface Tension Jar

Quickly and easily demonstrate the remarkable properties of surface tension and air pressure using a simple jar. Surface tension is a force—a force powerful enough to prevent water from spilling out of an open jar when it is turned upside-down! A fine mesh screen hidden inside the lid of the jar provides hundreds of tiny surface tension "membranes" that will support the weight of the water against the force of gravity. Look for the key "supporting" role played by the external air pressure.

Freezing by Boiling—Discrepant Event

The boiling point of a liquid depends on the external air pressure. When an aqueous acetone solution is placed under vacuum, the boiling point decreases and the liquid boils even at room temperature. Boiling, however, is an endothermic process. As the liquid boils, the temperature decreases, and the liquid soon freezes. Boiling and freezing occur simultaneously!

Wet Dry Ice—Triple Point of CO_2

From making fog to "boiling in water," dry ice is well-known for creating special effects. If dry ice is allowed to sublime in a sealed pipet bulb, the pressure due to carbon dioxide gas will increase to a point where the liquid form of carbon dioxide can be seen. The behavior of "wet dry ice" is an interesting way to demonstrate phase diagrams and the triple point.

Four-Square Diffusion—Vapor Pressure of Liquids and Solids

Finally—an amazing but simple way to demonstrate vapor pressure and diffusion. Simply sprinkle a few crystals of two different solids into opposite quadrants in a divided Petri dish, add a little acetone to a third quadrant, and observe. Within seconds, the solids begin to dissolve into puddles of liquid. The rate at which each solid dissolves illustrates how intermolecular forces influence the properties of liquids and solids.

Concepts

- Phase changes
- Heat of fusion
- Calorimetry

- Melting and freezing
- Evaporation and Condensation

- Surface tension
- Air pressure
- Adhesion vs. cohesion

- Boiling point
- Vapor pressure

- Phase changes
- Phase diagrams
- Triple point
- Sublimation

- Vapor pressure
- Diffusion
- Sublimation
- Intermolecular forces

Additional Activities

Please see the following experiments and demonstrations in other volumes of the *Flinn ChemTopic*™ *Labs* series for additional activities dealing with the properties of solids and liquids.

Elements, Compounds, and Mixtures, Volume 2

Simple Distillation—Separation of a Mixture

Distillation is a process for the purification or separation of the components in a liquid mixture. The mixture is heated to evaporate the volatile components, and the vapor is then condensed to a liquid. How does the composition of the distillate differ from the composition of the original liquid? Use this demonstration as an opportunity to discuss the role of science and technology in managing our natural resources—distillation is a key technology in desalination plants for the production of drinking water.

Chemical Bonding, Volume 5

Properties of Solids—Structure and Bonding

Looking for patterns in the properties of different substances can help students understand how and why atoms join together to form compounds. What kinds of forces hold atoms together? How does the nature of the forces holding atoms together influence the properties of a material? The purpose of this experiment is to study the physical properties of common solids and to investigate the relationship between the type of bonding in a substance and its properties.

Properties of Metals—Crystal Structure and Heat Treatment

Heat treatment of metals is used to increase their hardness and improve their "workability"—their ability to be bent and shaped. Annealing, hardening, and tempering produce remarkable changes in the properties of metals. Discover the effects of heating and cooling metals and correlate the changes with models of crystal structure with this interesting "bobby pin" activity.

Splatter Test—Properties of Liquids

The properties of liquids reflect the bonding within molecules and the nature and strength of forces between molecules. The "Splatter Test" demonstration will leave your students with a lasting impression of how intermolecular forces between molecules affect the rate of evaporation of a liquid.

Thermochemistry, Volume 10

Measuring Energy Changes—Heat of Fusion

Our everyday experience tells us that energy in the form of heat is needed to melt ice or boil water. What happens to the heat energy that is absorbed in these endothermic processes? Can the amount of heat energy be determined? The purpose of this experiment is to study the temperature and heat changes that occur when ice melts. Students first measure a heating curve for ice, water, and steam, and then use the heat equation to determine the amount of heat needed to melt ice.

Concepts

- Distillation
- Evaporation and condensation
- Boiling point

- Ionic bonding
- Covalent bonding
- Metallic bonding

- Crystal structure
- Body-centered cubic
- Face-centered cubic

- Intermolecular forces
- Hydrogen bonding
- Dipole–dipole interactions

- Heat vs. temperature
- Exothermic vs. endothermic
- Heat of fusion
- Enthalpy change

It's Just a Phase
Heating and Cooling Curves

Teacher Notes

Introduction

When freezing weather is predicted, Florida's orange growers spray their trees with water to prevent the fruit from freezing. As the water freezes, it releases heat to the surroundings and protects the fruit from damage. The temperature of the freezing water mixture will remain at the freezing point as long as both ice and water are present. Let's look at the temperature changes and the energy changes that take place when a liquid freezes or a solid melts.

Concepts

- Solids and liquids
- Phase changes
- Melting point
- Kinetic-molecular theory

Background

The temperature changes and energy changes that occur when a solid melts or a liquid freezes can best be understood by imagining what solids and liquids look like at the level of molecules or ions. Solids and liquids differ in how ordered or rigid their structures are and in the range of motion that the molecules or ions are allowed. Molecules in a crystalline solid are packed together in an ordered three-dimensional pattern, called the crystal lattice, where they are "held in place" by attractive forces between the molecules. The motion of molecules in the solid state is limited to vibrations (stretching and rocking motions)—the molecules are not free to move away from their fixed positions. The forces between molecules in the liquid state are less well understood. Molecules in the liquid state are free to move and are not locked in position. Attractive forces between molecules, however, tend to keep the molecules close together, so that their motion is perhaps best described as coordinated rather than independent.

A solid and its liquid are in equilibrium at the melting point, the temperature at which a crystalline solid becomes a liquid. The melting point of a pure substance is a characteristic physical property that can be used to identify a substance and to determine its purity. When a solid is heated, the temperature of the solid will increase until it reaches the melting point. Temperature is related to the average kinetic energy of the molecules—as the temperature increases, the average kinetic energy increases and the molecules begin to vibrate more rapidly. At the melting point, the vibrations become so rapid that the molecules begin to "break loose" from their fixed positions and melting occurs. *The temperature of the solid–liquid mixture will remain constant at the melting point until all the solid has melted.* Although the temperature remains constant at the melting point, heat must be added to break the attractive forces between molecules. In general, the more orderly the packing arrangement of molecules in the solid state and the stronger the attractive forces between molecules, the higher the melting point will be and the more heat that will be needed to melt the solid. The amount of heat energy required to melt a solid at its melting point is called the heat of fusion. The reverse process occurs when a liquid freezes. When a liquid freezes, energy in the form of heat is released to the surroundings. The same amount of heat required to melt a solid will be released by the liquid when it freezes or solidifies.

Most solid substances are crystalline solids, with atoms, ions or molecules arranged in an orderly repeating pattern. Crystalline solids tend to have sharp melting points. The discussion in the Background section is geared toward molecular solids, not ionic or network solids. Some solids, such as glass, rubber, and plastic, are amorphous—they lack an ordered internal structure. Amorphous solids do not have discrete melting points. The temperature at which an amorphous solid changes from a brittle solid state to a "fluid" or plastic state is called the glass transition temperature.

It's Just a Phase – Page 2

Experiment Overview

The purpose of this experiment is to investigate the solid–liquid phase changes for lauric acid, an organic compound that is used to make soap. Temperature versus time data will be measured as the melted solid is slowly cooled (Part A), and again as the fused solid is reheated (Part B). The data will be graphed and the resulting heating and cooling curves will be analyzed to determine whether the freezing point/melting point depends on the direction in which the physical change takes place.

Pre-Lab Questions

1. The *kinetic-molecular theory* (KMT) describes how close together the molecules are in a solid, liquid, and gas, their relative motion, and the attractive forces between the molecules. Use the KMT to explain the following properties of liquids and solids:

   (a) A liquid flows and takes the shape of its container.

   (b) Solids are generally incompressible.

   (c) Liquids have a definite volume.

   (d) A solid absorbs heat from its surroundings as it melts.

2. The following graph shows heating curve data for ice, water, and steam as heat energy is added to the system at a constant rate. (a) In what regions of the curve (A–E) is the average kinetic energy of the molecules increasing? (b) In what region of the curve are ice and water present together? (c) What happens to the heat energy that is absorbed by the molecules in the region where ice and water are both present?

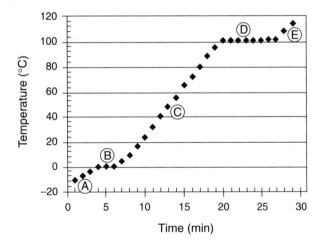

Teacher Notes

The heating curve data at the boiling point has been "condensed" in order to show all regions (A–E) of the graph. The temperature will theoretically remain constant about seven times longer at the boiling point than at the melting point. This is because ΔH_{vap} is seven times greater than ΔH_{fus}.

Materials

Lauric acid, $C_{11}H_{23}CO_2H$, 6 g
Beakers, 250- and 400-mL
Digital thermometers or temperature probes, 2*
Graph paper or computer graphing program
Hot plate or hot water bath*
Paper towels
Ring stand and clamp
Spatula
Styrofoam® cups, 2
Test tube, large, 18 × 150 mm
Test tube clamp or holder
Weighing dish

*One thermometer is used to monitor the temperature of the hot water bath. Groups may share hot water baths.

Teacher Notes

Safety Precautions

Read the entire procedure before beginning the experiment. Work carefully to avoid scalding yourself with hot water. Avoid contact of all chemicals with eyes and skin. Wash hands thoroughly with soap and water before leaving the laboratory. Wear chemical splash goggles, chemical-resistant gloves, and a chemical-resistant apron.

Procedure

Part A. Cooling Curve

1. Fill a 400-mL beaker two-thirds full with hot tap water. Heat the water to about 80 °C on a hot plate. Proceed to steps 2–5 as the water is heating. *Note:* After the lauric acid test tube has been removed from the hot water bath (step 5), lower the setting on the hot plate and reduce the temperature of the bath to about 60 °C.

2. Obtain about 6 g of lauric acid in a weighing dish and transfer the solid to a clean and dry test tube.

3. Add about 100 mL of cold tap water (15–20 °C) to a Styrofoam cup and nest the cup inside a second Styrofoam cup. Place the nested Styrofoam cups in a 250-mL beaker and set the beaker on the ring stand.

4. Holding the test tube (prepared in step 2) with a clamp or test tube holder, place the lauric acid into the hot water bath (step 1) at 80 °C.

5. Insert a digital thermometer into the lauric acid. When the temperature is about 75 °C, remove the test tube from the hot water bath and clamp the test tube to the ring stand.

6. Measure and record the precise temperature of the melted lauric acid in the test tube and then *immediately* lower the test tube into the cold water bath in the Styrofoam cup. *Start timing.*

7. Carefully stir the lauric acid with the digital thermometer and measure the temperature every *30 seconds* for 10 minutes, or until the temperature is about 30 °C (whichever comes first). Record all time and temperature readings in the data table. *Note:* Continue stirring the sample until it is no longer possible to do so (when the lauric acid has solidified).

Part B. Heating Curve

8. Check that the temperature of the hot water bath is between 58 and 60 °C. Do not let the temperature rise above 60 °C—add cold water if necessary to adjust the temperature.

9. Remove the test tube from the cold water bath (step 7). Measure and record the precise temperature of the solid and *immediately* place the test tube in the 60 °C water bath. *Start timing.*

10. Measure and record the temperature of the lauric acid every *30 seconds* for 10 minutes, or until the temperature is about 55 °C (whichever comes first). *Note:* Start stirring once the lauric acid has softened up enough to do so, but do not try to "force" it.

11. Remove the test tube from the hot water bath and dispose of the lauric acid as directed by the instructor.

Do not use a glass thermometer to stir the sample in the test tube. Heating a water bath is always a slow step in experiments. To save time, have pre-melted lauric acid samples in ready-to-use test tubes in one central hot water bath in the front of the room. Large coffee urns or lab microwaves are an excellent way to quickly heat water for hot water baths. If there is a microwave oven in the lab, make sure that it is designated for lab use only.

Name: _____

Class/Lab Period: _____

It's Just a Phase

Data Table

Cooling Curve (Part A)		Heating Curve (Part B)	
Time	Temperature (°C)	Time	Temperature (°C)
0 sec		0 sec	
30 sec		30 sec	
1 min		1 min	
1 min 30 sec		1 min 30 sec	
2 min		2 min	
2 min 30 sec		2 min 30 sec	
3 min		3 min	
3 min 30 sec		3 min 30 sec	
4 min		4 min	
4 min 30 sec		4 min 30 sec	
5 min		5 min	
5 min 30 sec		5 min 30 sec	
6 min		6 min	
6 min 30 sec		6 min 30 sec	
7 min		7 min	
7 min 30 sec		7 min 30 sec	
8 min		8 min	
8 min 30 sec		8 min 30 sec	
9 min		9 min	
9 min 30 sec		9 min 30 sec	
10 min		10 min	

Teacher Notes

Post-Lab Questions

1. Prepare a graph of temperature on the y-axis versus time on the x-axis. Plot the data from Parts A and B as two series of points, using different color pencils or different shapes to mark the points for Part A versus Part B. Draw a smooth (continuous) curve through the plotted points for each series of data.

2. Label the following regions (A–C) on the *cooling curve:* A, only liquid is present; B, liquid and solid are present together; C, only solid is present.

3. What happens to the temperature of a substance while it is freezing or melting? Estimate the freezing point and the melting point of lauric acid from the cooling curve and the heating curve, respectively. Does the freezing point/melting point depend on the direction in which the phase change takes place?

4. Which set of data (the cooling curve or the heating curve) provided a more accurate or a more precise estimate of the melting point? What variables in the design of the experiment might account for any difference in the results?

5. *Circle the correct choices:* Freezing is an (exothermic/endothermic) process—the liquid (absorbs/releases) heat from or to its surroundings. At the freezing point, the average (kinetic energy/potential energy) of the molecules (increases/decreases) and the liquid solidifies.

6. Answer *increases, decreases,* or *no change* to predict how doubling the amount of lauric acid would change the results in Part B:

   The *rate* at which the temperature of the solid increases.   _____

   The *temperature* at which the solid melts.   _____

   The *amount of heat* absorbed by the lauric acid as it melts.   _____

7. *(Optional)* Lauric acid is a fatty acid (a component of fats and oils). What factor might explain the regular increase in the melting point of the following fatty acids as the number of carbon atoms increases?

   Lauric acid, $CH_3(CH_2)_{10}CO_2H$, mp 43.2 °C

   Myristic acid, $CH_3(CH_2)_{12}CO_2H$, mp 54.0 °C

   Palmitic acid, $CH_3(CH_2)_{14}CO_2H$, mp 61.8 °C

   Stearic acid, $CH_3(CH_2)_{16}CO_2H$, mp 68.8 °C

Teacher's Notes
It's Just a Phase

Master Materials List *(for a class of 30 students working in pairs)*

Lauric acid or organic "unknown," 90 g
Beakers, 250- and 400-mL, 15 each
Digital thermometers or temperature sensors, 22*
Graph paper or computer graphing program
Hot plates, 7, or hot water baths*
Paper towels

Ring stands and clamps, 15
Spatulas, 15
Styrofoam cups, 30
Test tubes, Pyrex®, large, 18 × 150 mm, 15
Test tube clamps or holders, 15
Weighing dishes, 15

Two groups may share one hot plate or hot water bath.

Safety Precautions

Use only borosilicate (e.g., Pyrex®) test tubes for this laboratory. Inspect all test tubes and do not use any cracked or chipped test tubes. Warn students to work carefully to avoid scalding themselves with the hot water bath. Avoid contact of all chemicals with eyes and skin. Wear chemical splash goggles, chemical-resistant gloves, and a chemical-resistant apron. Remind students to wash their hands thoroughly with soap and water before leaving the laboratory. Please review current Material Safety Data Sheets for additional safety, handling, and disposal information.

Disposal

Please consult your current *Flinn Scientific Catalog/Reference Manual* for general guidelines and specific procedures governing the disposal of laboratory waste. Lauric acid may be disposed of according to Flinn Suggested Disposal Method #24a. Alternatively, the lauric acid samples may be recycled from class to class and also from year to year. To recycle the samples from one lab period to the next, simply collect the test tubes at the end of the period and place them in a central hot water bath—the test tubes will be ready to use in Part A, step 6. To recycle the lauric acid samples for future use, stopper the test tubes and store them in a labeled container.

Lab Hints

- The laboratory work for this experiment can easily be completed in a typical 50-minute class period. If desired, however, the lab may be simplified by dividing the class into groups of four students, with two "working pairs" in each group. Each working pair would be responsible for only one part of the experiment, either Part A or Part B. The two working pairs would then share the data with each other to complete the *Post-Lab Questions*. This approach may also be used if the number of hot plates is a limiting factor. Two groups may share one hot plate or hot water bath. Save time by pre-melting the lauric acid samples in a central hot water bath.

Teacher's Notes

Teacher Notes

- Excellent cooling curve data is obtained in Part A with as little as 2–3 g of lauric acid in a medium test tube (16 × 125 mm). In order to get good results in Part B, however, a larger sample size (6 g of lauric acid) is recommended. When smaller sample sizes were tested in Part B, the temperature of the solidified lauric acid steadily increased when placed in a 60 °C water bath and the heating curve did not exhibit a well-defined "plateau" at the melting point. Three variables—the amount of lauric acid, the temperature of the hot water bath, and the amount of water in the hot water bath—will affect the quality of the heating curve data.

- This experiment may also be carried out using solid, low-melting organic "unknowns" for the students to identify. Suitable unknowns that may be used in addition to lauric acid include cetyl alcohol (1-hexadecanol, $C_{16}H_{33}OH$, mp 54–56 °C), stearic acid (octadecanoic acid, $C_{17}H_{35}CO_2H$, mp 67–69 °C), and BHT (butylated hydroxytoluene, 2,6-di-t-butyl-4-methylphenol, $C_{15}H_{23}OH$, mp 69–71 °C). Phenyl salicylate ("salol," $HOC_6H_4CO_2C_6H_5$, mp 44–46 °C) has a melting point close to that of lauric acid. Acetamide, a traditional favorite for melting point and heat of fusion experiments, is classified as a possible carcinogen and is not recommended.

- The experiment may be extended to determine the heat of fusion of lauric acid. See "Hot Wax" in the *Demonstrations* section of this book for a sample calorimetry procedure and results. The heat of fusion procedure gives good results but requires a larger sample size (about 10 g of lauric acid), which increases the time needed to complete Parts A and B. This extension may be more suitable for an honors or advanced class. It is not recommended as a general procedure for all classes.

Teaching Tips

- The properties of solids and liquids and phase changes have been the subject(s) of experiments in previous volumes of the *Flinn ChemTopic™ Labs* series. In "Properties of Solids" (see *Chemical Bonding,* Volume 5 in the series), students compare the general properties of ionic, molecular, covalent-network, and metallic solids. In "Measuring Energy Changes" (see *Thermochemistry,* Volume 10 in the series), students obtain heating curve data for ice, water, and steam and then measured the heat of fusion of water.

- "Phase-change wallboard" has been developed as a passive thermal storage construction material to improve energy efficiency in heating and cooling buildings. Phase-change wallboard contains paraffin wax embedded in gypsum. Paraffin is used as a "phase change material" (PCM); it undergoes reversible phase changes when heated or cooled, absorbing or releasing large amounts of heat and maintaining a constant temperature in the process. Consider how a PCM might be used in buildings in a mild-climate area. During the day, when the outside temperature increases, the solid PCM melts and absorbs (stores) the excess heat energy. This cools the building and reduces the need for external cooling (air conditioning). The reverse occurs at night—when the outside temperature decreases, the liquid PCM solidifies and releases its "stored" heat energy.

Teacher's Notes

Answers to Pre-Lab Questions *(Student answers will vary.)*

1. The *kinetic-molecular theory* (KMT) describes how close together the molecules are in a solid, liquid, and a gas, their relative motion, and the attractive forces between the molecules. Use the KMT to explain the following properties of liquids and solids:

   (a) A liquid flows and takes the shape of its container.

   (b) Solids are generally incompressible.

   (c) Liquids have a definite volume.

   (d) A solid absorbs heat from its surroundings as it melts.

   (a) The molecules in a liquid move freely and randomly.

   (b) The molecules in a solid are physically as close together as possible (tightly packed).

   (c) The molecules in a liquid are close together and there are attractive forces between the molecules.

   (d) Relatively strong attractive forces exist between the molecules in the solid state—these forces must be (partially) broken when the solid changes to a liquid.

2. The following graph shows heating curve data for ice, water, and steam as heat energy is added to the system at a constant rate. (a) In what regions of the curve (A–E) is the average kinetic energy of the molecules increasing? (b) In what region of the curve are ice and water present together? (c) What happens to the heat energy that is absorbed by the molecules in this region of the curve?

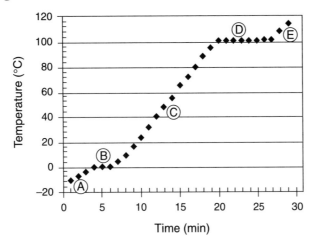

   (a) The average kinetic energy of the molecules increases as the temperature increases in regions A, C, and E.

   (b) Ice and water are present together at the melting point of water (0 °C, region B).

   (c) The heat energy absorbed by the ice molecules is used to break the attractive forces between the molecules in the crystal lattice. **Note to teachers:** *The average kinetic energy of the molecules is constant. The heat energy increases the potential energy of the molecules. At the same temperature, molecules in the liquid state have a higher potential energy than molecules in the solid state.*

Teacher's Notes

Teacher Notes

Sample Data

Student data will vary.

Data Table

Cooling Curve (Part A)		Heating Curve (Part B)	
Time	**Temperature (°C)**	**Time**	**Temperature (°C)**
0 sec	65.8	0 sec	22.9
30 sec	50.6	30 sec	24.6
1 min	45.1	1 min	30.4
1 min 30 sec	44.2	1 min 30 sec	34.9
2 min	44.0	2 min	38.2
2 min 30 sec	44.0	2 min 30 sec	40.4
3 min	44.0	3 min	42.0
3 min 30 sec	44.0	3 min 30 sec	42.9
4 min	44.0	4 min	43.3
4 min 30 sec	44.0	4 min 30 sec	43.7
5 min	43.9	5 min	44.4
5 min 30 sec	43.6	5 min 30 sec	45.6
6 min	42.8	6 min	46.6
6 min 30 sec	39.6	6 min 30 sec	49.3
7 min	34.0	7 min	53.2
7 min 30 sec	28.7*	7 min 30 sec	55.6*
8 min		8 min	
8 min 30 sec		8 min 30 sec	
9 min		9 min	
9 min 30 sec		9 min 30 sec	
10 min		10 min	

*The cooling and heating curve runs may be stopped when the temperature reaches 30 °C and 55 °C, respectively.

It's Just a Phase

Teacher's Notes

Answers to Post-Lab Questions *(Student answers will vary.)*

1. Prepare a graph of temperature on the *y*-axis versus time on the *x*-axis. Plot the data from Parts A and B as two series of points, using different color pencils or different shapes to mark the points for Part A versus Part B. Draw a smooth (continuous) curve through the plotted points for each series of data.

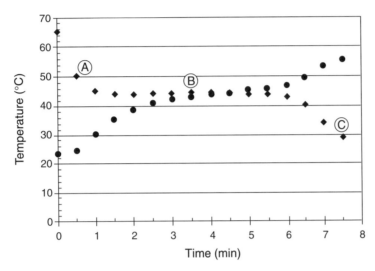

2. Label the following regions (A–C) on the *cooling curve:* A, only liquid is present; B, liquid and solid are present together; C, only solid is present.

3. What happens to the temperature of a pure substance while it is freezing or melting? Estimate the freezing point and the melting point of lauric acid from the cooling curve and the heating curve, respectively. Does the freezing point/melting point depend on the direction in which the phase change takes place?

    The temperature of a pure substance should remain constant at the freezing point or melting point as long as both phases are present. The freezing point of lauric acid is estimated from the "flat" region of the cooling curve—44.0 °C. The heating curve did not "level off" at a particular temperature as did the cooling curve. However, the rate at which the temperature increased slowed down noticeably in the 43–44 °C temperature range, giving an estimate of 43.5 °C for the melting point. The slight difference in the estimated freezing point/melting point is not significant (within experimental error). The freezing point/melting point does not depend on the direction in which the phase change takes place.

4. Which set of data (the cooling curve or the heating curve) provided a more accurate or a more precise estimate of the melting point? Which variables in the design of the experiment might account for any difference in the results?

    The literature melting point of lauric acid is 43.2 °C (see Question #7). The heating curve data gave a more accurate estimate of the melting point, but the cooling curve data gave a more precise value. The cooling curve was qualitatively easier to interpret because there was a true temperature plateau at the freezing point. The amount of lauric acid was the same in both parts of the experiment. Several variables, however, were different in Part A versus Part B. The hot water bath contained

Teacher's Notes

more water than the cold water bath and was therefore a greater heat "reservoir." The solid sample in Part B could not be stirred, making heat transfer less efficient. The cold water bath was insulated in a coffee cup, the hot water bath was not. **Note to teachers:** *This is a great discussion question—students usually grasp at straws when it comes to analyzing experimental error. Ask students to consider how these variables would affect the rate at which heat is transferred into or out of the system, and how the temperature versus time data would be affected as a result.*

5. *Circle the correct choices:* Freezing is an **exothermic** process—the liquid **releases** heat to its surroundings. At the freezing point, the average **potential energy** of the molecules **decreases** and the liquid solidifies.

    Note to teachers: *Question #5 is difficult for students to grasp. Energy is released during freezing but the temperature does not change. Average kinetic energy is dependent on temperature so it does not change. The potential energy must decrease and this is due to the formation of intermolecular forces holding the solid together. Molecules in the liquid state have a higher potential energy than molecules in the solid state.*

6. Answer *increases, decreases,* or *no change* to predict how doubling the amount of lauric acid would change the results in Part B:

The *rate* at which the temperature of the solid increases.	*Decreases*
The *temperature* at which the solid melts.	*No change*
The *amount of heat* absorbed by the lauric acid as it melts.	*Increases*

7. *(Optional)* Lauric acid is a fatty acid (a component of fats and oils). What factor might explain the regular increase in the melting point of the following fatty acids as the number of carbon atoms increases?

    Lauric acid, $CH_3(CH_2)_{10}CO_2H$, mp 43.2 °C

    Myristic acid, $CH_3(CH_2)_{12}CO_2H$, mp 54.0 °C

    Palmitic acid, $CH_3(CH_2)_{14}CO_2H$, mp 61.8 °C

    Stearic acid, $CH_3(CH_2)_{16}CO_2H$, mp 68.8 °C

    In general, the greater the attractive forces between the molecules in a molecular solid, the higher its melting point. This implies that the strength of the attractive forces increases as the number of carbon atoms increases. **Note to teachers:** *The properties of the fatty acids are dominated by the long nonpolar hydrocarbon "tail." The principal attractive forces acting between nonpolar molecules are London dispersion forces. The strength of London dispersion forces increases as the size of the molecules increases.*

Teacher's Notes

Teacher Notes

Teacher Notes

Properties of Liquids
Surface Tension and Capillary Action

Introduction

Have you ever seen an insect or spider appear to "walk" on water? The ability of water bugs to stay on top of the water is due to its very high surface tension, which acts like an invisible film that prevents the bug from breaking the surface. Surface tension and other properties of liquids depend on the nature and the strength of attractive forces between molecules.

Concepts

- Properties of liquids
- Surface tension
- Capillary action
- Intermolecular forces

Background

The properties of liquids are due to the motion of molecules in the liquid phase and the existence of attractive forces between molecules. According to the kinetic-molecular theory, the molecules in a liquid are in constant, random motion. The molecules are close enough together, however, that attractive forces between neighboring molecules influence their motion and give liquids their characteristic properties. Comparing the properties of different liquids allows us to compare the strength of attractive forces between different types of molecules. The boiling point of a liquid, for example, reflects the ability of molecules in the liquid phase to break the attractive forces between them and "escape" into the vapor phase. Liquids with low boiling points are considered volatile—they will evaporate readily from an open container. The boiling point of a liquid depends on the strength of attractive forces between the molecules. Liquids with strong intermolecular attractive forces have higher boiling points than liquids with weaker attractive forces.

Surface tension is a result of uneven attractive forces experienced by molecules at the surface versus those in the rest of the liquid (Figure 1). Molecules in the liquid are bound to neighboring molecules all around them. Molecules at the surface, however, have no neighboring molecules above them. Because the forces acting on the surface molecules are not balanced in all directions, the surface molecules are drawn inward toward the rest of the liquid. The result is *surface tension*—a net attractive force that tends to pull adjacent surface molecules inward, thus decreasing the surface area to the smallest possible size. Surface tension acts as an invisible film that makes it more difficult to move an object through the surface than through the bulk of a liquid.

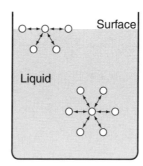

Figure 1. Attractive Forces Between Molecules and the Origin of Surface Tension.

Properties of Liquids – Page 2

The cause of surface tension (uneven attractive forces) explains why water has a very high surface tension compared to that of other liquids. The greater the forces of attraction between molecules in the liquid phase, the higher the surface tension will be. There are three fundamental types of intermolecular attractive forces—London dispersion forces, dipole-dipole interactions, and hydrogen bonding. Nonpolar molecules are attracted to each other by relatively weak and fleeting London dispersion forces. Polar molecules that contain permanent dipoles are bound by stronger electrostatic forces, called dipole-dipole interactions. Molecules containing highly polar O–H or N–H groups form hydrogen-bonded networks with adjacent molecules. Water's high surface tension is a consequence of its strong hydrogen bonds and explains such phenomena as how water striders walk on water and why water beads up into large droplets on a waxy leaf surface.

Capillary action, the rise of liquid in a narrow tube or fiber, is another natural phenomenon that may be explained in terms of surface tension. Examples of capillary action include the transport of water from the roots of a plant to its leaves and the migration of solvent in paper chromatography. A liquid will rise quite high in a narrow tube if there are strong attractive forces (adhesion) between the liquid molecules and the molecules that make up the surface of the tube. Adhesion tends to pull the liquid molecules upward along the surface of the tube, against the force of gravity. The surface tension acts to hold the surface together, so instead of just the edges moving upward, the whole liquid surface is dragged upward. The height to which capillary action will lift a liquid depends on the weight of the liquid which the surface tension will support—narrow tubes and high surface tension result in tall columns of liquid. The overall rise of a liquid in a capillary tube is proportional to the surface tension and inversely proportional to the density of the liquid.

Experiment Overview

The purpose of this experiment is to observe the effect of surface tension in water and then to compare the surface tension and capillary action of different liquids.

Pre-Lab Questions

1. Draw the structure of a water molecule and show by means of a diagram the hydrogen bonds between water molecules. How many hydrogen bonds does each water molecule form?

2. Hydrocarbons are nonpolar compounds containing carbon and hydrogen atoms. The properties of three hydrocarbons are summarized below. (a) How do the attractive forces between molecules change in the transition from the gas to the liquid to the solid state? (b) Based on its properties, which compound has the strongest attractive forces? The weakest attractive forces? (c) Write a general statement describing how the size of a molecule influences the strength of London dispersion forces between molecules.

Methane	Octane	Eicosane
CH_4	$CH_3CH_2CH_2CH_2CH_2CH_2CH_2CH_3$	$CH_3(CH_2)_{18}CH_3$
Natural gas	Gasoline	Lubricant (grease)
Gas, bp –161 °C	Liquid, bp 126 °C	Solid, mp 37 °C

Teacher Notes

*The relative strength of intermolecular forces versus covalent bonds is often confusing to students. Because we tend to emphasize **very strong** hydrogen bonds, students begin to think of hydrogen bonds as stronger than covalent bonds. This is not true! Hydrogen bonds are very strong attractive forces between molecules. The strongest hydrogen bond, however, is probably only 5% the strength of an average covalent bond.*

Page 3 – **Properties of Liquids**

Teacher Notes

Materials

Distilled water, 1 mL

Tap water

Isopropyl alcohol, $(CH_3)_2CHOH$, 1 mL

25% Isopropyl alcohol, 1 mL

70% Isopropyl alcohol, 1 mL

Soap solution (sodium lauryl sulfate), 1%, 1 mL

Glass jar or beaker, 125- or 250-mL

Beral-type pipets, 6

Capillary tubes, open-ended, 6

Metric ruler

Paper towels

Pennies, 30–40

Reaction strip, 8- or 12-well

Safety Precautions

Isopropyl alcohol is a flammable liquid and a dangerous fire risk—avoid contact with flames and heat. It is slightly toxic by ingestion and inhalation. Avoid contact of all chemicals with eyes and skin. Wear chemical splash goggles, chemical-resistant gloves, and a chemical-resistant apron. Wash hands thoroughly with soap and water before leaving the laboratory.

Procedure

Part A. Surface Tension Demonstration

1. Do this part in a group with four students. Place a small glass jar on a mat of paper towels and fill the jar with tap water all the way to the top until the water just begins to overflow.

2. Predict how many pennies can be added to the jar without causing the water to spill over the edge. Record your group's estimate in the data table and give a reason for the "guess."

3. Carefully add a penny to the jar without spilling any water. *Hint:* This may take some practice. Gently place the penny halfway through the water surface and then let it fall slowly to the bottom.

4. Continue adding pennies one at a time to the jar. What happens to the top of the water surface as more and more pennies are added? Describe your observations in the data table.

5. Record how many pennies were added to the jar before water began to spill over the edge in the data table.

Part B. Capillary Action

6. Obtain an 8- or 12-well reaction strip. Fill well #1 about two-thirds full with distilled water.

7. Place an open-ended capillary tube in the well and allow the water level in the tube to stabilize (about 15 seconds). Carefully remove the capillary tube from the well and measure the height in mm of the liquid in the tube. Record the liquid height in the data table.

8. Remove the capillary tube from the well and immediately blot the tube on a paper towel to drain the liquid. *Note:* Do not touch the bottom of the capillary tube with your hands. Even trace amounts of skin oils will change the surface tension of water.

Skin oils will interfere with capillary action in Part B. It takes only a very small amount of a nonpolar substance to drastically reduce the surface tension of water. Instruct students to handle the capillary tubes only from the top end.

Properties of Liquids

Properties of Liquids – Page 4

9. Repeat steps 7 and 8 two more times using the same liquid to obtain a total of three measurements. Use a fresh capillary tube if air bubbles develop in the tube.

10. Fill well #2 about two-thirds full with isopropyl alcohol. Using a fresh capillary tube, repeat steps 7–9 to measure the rise in liquid level due to capillary action. Record all three measurements in the data table.

11. Fill well #3 about two-thirds full with 25% isopropyl alcohol. Using a fresh capillary tube, repeat steps 7–9 to measure the rise in liquid level due to capillary action.

12. Fill well #4 about two-thirds full with 70% isopropyl alcohol. Using a fresh capillary tube, repeat steps 7–9 to measure the rise in liquid level due to capillary action.

13. Fill well #5 about two-thirds full with tap water. Using a fresh capillary tube, repeat steps 7–9 to measure the rise in liquid level due to capillary action.

14. Fill well #6 about two-thirds full with 1% soap solution. Using a fresh capillary tube, repeat steps 7–9 to measure the rise in liquid level due to capillary action.

Teacher Notes

It is important not to contaminate any of the wells or capillary tubes with soap (step 14). Always do soap last!

Teacher Notes

Name: _____

Class/Lab Period: _____

Properties of Liquids

Data Table

Part A. Surface Tension Demonstration	
How many pennies may be added to the jar before the water overflows?	Prediction: Explain: Observations: Number of pennies added (step 5):

Part B. Capillary Action	Height of Liquid Column (mm)			
Liquid	Trial 1	Trial 2	Trial 3	Average
Distilled Water				
Isopropyl Alcohol				
25% Isopropyl Alcohol				
70% Isopropyl Alcohol				
Tap Water				
1% Soap Solution				

Post-Lab Questions

1. Explain why the water in an apparently full glass or jar does not overflow as pennies are added to the water. What force prevents the water from spilling over the edge?

2. Based on capillary action, which liquid has a higher surface tension, water or isopropyl alcohol? Explain in terms of the polarity of water versus isopropyl alcohol and the strength of their attractive forces.

Review the structures of water and isopropyl alcohol molecules, as well as the similarities and differences between them.

Properties of Liquids

Properties of Liquids – Page 6

3. Plot the data for pure water, pure isopropyl alcohol, and isopropyl alcohol–water solutions on the following graph. Note that the *x*-axis is *percent water* (not percent isopropyl alcohol).

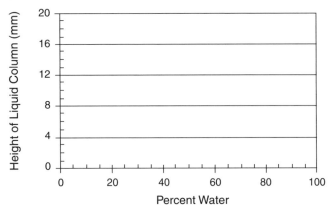

4. Which statement better describes the observed trend? (a) The change in surface tension is proportional to the amount of isopropyl alcohol added to water; OR (b) Adding even a small amount of isopropyl alcohol produces a large change in the surface tension of water. Suggest a reason for the observed trend.

5. What effect does soap have on the surface tension of water? How is this related to how soaps work?

6. Tap water contains dissolved solids, in particular, magnesium and calcium salts. How do these salts change the surface tension of water?

7. *(Optional)* Soaps and detergents work better in hot water than in cold water. What effect should increasing the temperature have on the surface tension of water? Explain in terms of the kinetic-molecular theory and the attractive forces between molecules.

Teacher's Notes
Properties of Liquids

Master Materials List *(for a class of 30 students working in pairs)*

Distilled water, 15 mL	Beral-type pipets, 6–90 (see *Hints*)
Tap water	Capillary tubes, open-ended, 90
Isopropyl alcohol, $(CH_3)_2CHOH$, 15 mL	Metric rulers, 15
25% Isopropyl alcohol, 15 mL	Paper towels
70% Isopropyl alcohol, 15 mL	Pennies, about 300*
Soap solution (sodium lauryl sulfate), 1%, 15 mL	Reaction strips, 8- or 12-well, 15
Small glass jars or beakers, 7*	

*Small (approx. 4- or 8-oz.) battery jars or "specimen jars" work better than beakers with spouts. Baby food jars may be convenient. Ask students to bring in their own pennies!

Safety Precautions

Isopropyl alcohol is a flammable liquid and a dangerous fire risk—avoid contact with flames and heat. It is slightly toxic by ingestion and inhalation. Avoid contact of all chemicals with eyes and skin. Wear chemical splash goggles, chemical-resistant gloves, and a chemical-resistant apron. Please review current Material Safety Data Sheets for additional safety, handling, and disposal information. Remind students to wash hands thoroughly with soap and water before leaving the laboratory.

Disposal

Please consult your current *Flinn Scientific Catalog/Reference Manual* for general guidelines and specific procedures governing the disposal of laboratory waste. Isopropyl alcohol solutions may be rinsed down the drain with plenty of water according to Flinn Suggested Disposal Method #26b.

Lab Hints

- The laboratory work for this experiment can reasonably be completed in about 20–30 minutes. This allows ample time within a typical classroom period to review the background material, answer the *Pre-Lab Questions,* make predictions, and discuss the results. If time permits, use the "Surface Tension Jar" demonstration in this book to prove that surface tension is a real force—stronger than gravity!

- If each group takes a Beral-type pipet of solution back to their work area, about 90 pipets will be needed. If each group brings their well plate to the solution bottles, then only 6 or 12 pipets will be needed. Rubberband test tubes to reagent bottles so students can place used pipets in an uncontaminated storage site.

- Surface tension is not the only variable in Part B. Capillary action is directly proportional to surface tension and inversely proportional to density. The density of isopropyl alcohol is 0.79 g/mL. If isopropyl alcohol and water had the same surface tension, then the liquid rise would be 25% *higher* for isopropyl alcohol than for water. Instead, the liquid rise is only about one-third that of water. At room temperature, the surface tension of water is 0.073 N/m and the surface tension of isopropyl alcohol is 0.024 N/m (where N is Newtons).

Teacher's Notes

- Many different liquids may be tested in Part B. To avoid confusion, however, try not to introduce too many variables in terms of the properties of the liquids that are being compared. The series of aliphatic alcohols (methyl alcohol, ethyl alcohol, propyl alcohol, etc.) provides an interesting and perhaps counter-intuitive comparison. All of the alcohols have about the same density (0.79 g/mL). They also have almost identical surface tensions—0.022–0.024 N/m. This suggests that differences in surface tension are not as "simple" as differences in attractive forces. Another good series for comparison would be isopropyl alcohol, propylene glycol, and glycerol. These compounds have the same number of carbon atoms but one, two, and three hydroxyl groups, respectively. Correlating their capillary action results, however, is complicated by the fact that they have very different densities (d = 0.79, 1.1, and 1.26 g/mL for isopropyl alcohol, propylene glycol, and glycerol, respectively). The fact that glycerol is much more viscous than water might lead students to conclude that capillary action and surface tension depend on or are related to viscosity. Surface tension and viscosity are, in fact, independent properties of a liquid.

- The effect of temperature on the surface tension of water is an obvious extension for Part B. You may also encourage students to suggest other conditions or additives that will affect the surface tension of water.

Teaching Tips

- Surface tension is related to disinfectant activity for alcohols and soaps as well as other "surface-active agents." Low surface tension allows these liquids to spread out on and disrupt the cell walls or cell membrane of microorganisms. [Alcohols also denature or coagulate the proteins in the cell membrane. Rubbing alcohol (70% isopropyl alcohol) is thus more effective than absolute alcohol.]

- The experiment "How Cool Is That?—Evaporation of Liquids" on page 35 in this book is a complementary lab for studying the properties of liquids. Students measure temperature-versus-time cooling curves as different liquids evaporate and compare the minimum temperature reached for each liquid. Pair-wise comparisons (e.g., methyl alcohol versus ethyl alcohol, acetone versus isopropyl alcohol, hexane versus heptane) help students relate the rate of evaporation of liquids to the nature and strength of attractive forces between molecules.

- See the "Splatter Test" demonstration in *Chemical Bonding,* Volume 5 in the *Flinn ChemTopic™ Labs* series for a simple but effective comparison of the rate of evaporation of organic solvents (acetone, ethyl alcohol, and cyclohexane).

Teacher's Notes

Teacher Notes

Answers to Pre-Lab Questions *(Student answers will vary.)*

1. Draw the structure of a water molecule and show by means of a diagram the hydrogen bonds between water molecules. How many hydrogen bonds does each water molecule form?

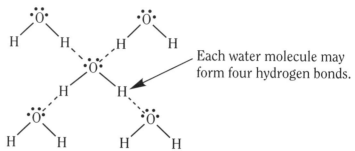

Each water molecule may form four hydrogen bonds.

2. Hydrocarbons are nonpolar compounds containing carbon and hydrogen atoms. The properties of three hydrocarbons are summarized below. (a) How do the attractive forces between molecules change in the transition from the gas to the liquid to the solid state? (b) Based on its properties, which compound has the strongest attractive forces? The weakest attractive forces? (c) Write a general statement describing how the size of a molecule influences the strength of London dispersion forces between molecules.

Methane	**Octane**	**Eicosane**
CH_4	$CH_3CH_2CH_2CH_2CH_2CH_2CH_2CH_3$	$CH_3(CH_2)_{18}CH_3$
Natural gas	Gasoline	Lubricant (grease)
Gas, bp –161 °C	Liquid, bp 126 °C	Solid, mp 37 °C

(a) *Attractive forces between molecules increase in the order gas << liquid < solid. Molecules in the gas state are very far apart—there are almost no attractive forces between the molecules. As gases condense into liquids and then solidify, the molecules get closer together and the strength of attractive forces between molecules increases. Attractive forces between molecules are strongest in the solid state, because the molecules are locked into fixed positions.*

(b) *Eicosane, a hydrocarbon with a chain of 20 carbon atoms, has stronger intermolecular attractive forces than octane or methane, which contain eight carbon atoms and one carbon atom, respectively. Methane has the weakest attractive forces.*

(c) *The strength of London dispersion forces between molecules increases as the size of the molecules increases (all other factors being equal).*

Properties of Liquids

Teacher's Notes

Sample Data

Student data will vary.

Data Table

Part A. Surface Tension Demonstration	
How many pennies may be added to the jar before the water overflows?	**Prediction:** The jar looks pretty full! Maybe 5–10 pennies—enough to form one layer on the bottom of the jar? **Explain:**
	Observations: The water surface at the top of the jar becomes rounded, like a dome. The surface extends about 5 mm over the top of the beaker. **Number of pennies added (step 5):** > 40!*

Part B. Capillary Action	Height of Liquid Column (mm)			
Liquid	Trial 1	Trial 2	Trial 3	Average
Distilled Water	17	17	18	17
Isopropyl Alcohol	8	7	9	8
25% Isopropyl Alcohol	10	10	11	10
70% Isopropyl Alcohol	8	9	8	8
Tap Water	18	18	18	18
1% Soap Solution	11	11	10	11

In a 250-mL glass jar.

Answers to Post-Lab Questions *(Student answers will vary.)*

1. Explain why the water in an apparently full glass or jar does not overflow as pennies are added to the water. What force prevents the water from spilling over the edge?

   Surface tension is a force! The force essentially "pulls" the molecules on the surface toward the interior volume of the liquid. Surface tension prevents the water from spilling over the edge even as the total volume of the pennies and the water exceeds the capacity of the beaker. **Note to teachers:** *A good analogy here is that the water appears to have an elastic surface film, similar to that of a balloon, which expands as needed to accommodate the added volume.*

Teacher's Notes

Teacher Notes

2. Based on capillary action, which liquid has a higher surface tension, water or isopropyl alcohol? Explain in terms of the polarity of water versus isopropyl alcohol and the strength of their attractive forces.

   Water has a significantly higher surface tension than isopropyl alcohol. Water is also a more polar liquid than isopropyl alcohol and has more extensive hydrogen bonding because it has two O—H bonds. (The nonpolar C_3H_7- group in isopropyl alcohol also decreases its overall hydrogen-bonding tendency.) Surface tension thus appears to correlate with the attractive forces between the molecules in water versus isopropyl alcohol.

3. Plot the data for pure water, pure isopropyl alcohol, and isopropyl alcohol–water solutions on the following graph. Note that the *x*-axis is *percent water* (not percent isopropyl alcohol).

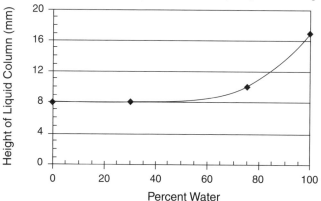

4. Which statement better describes the observed trend? (a) The change in surface tension is proportional to the amount of isopropyl alcohol added to water; OR (b) Adding even a small amount of isopropyl alcohol produces a large change in the surface tension of water. Suggest a reason for the observed trend.

   Statement (b) more accurately describes the observed trend. The graph is definitely not linear. It appears that even a small amount of isopropyl alcohol is sufficient to disrupt the extensive hydrogen-bonded network in the structure of liquid water. **Note to teachers:** *Water truly is an exceptional liquid! Remind students that ice is less dense than water—the "fluid" association of water molecules is somehow greater in the liquid state than in the solid state.*

5. What effect does soap have on the surface tension of water? How is this related to how soap works?

   Soap dramatically lowers the surface tension of water. One way that soaps work is by improving the "wettability" of surfaces (skin, clothing, etc.). Water's extremely high surface tension prevents it from spreading out across and penetrating into a material. Soap lowers the surface tension of water and helps it diffuse into the pores on these surfaces. **Note to teachers:** *If time permits, show the structure of a typical "liquid soap," such as sodium lauryl sulfate (see "It's Just a Phase"). All soaps have two things in common—an ionic or polar (hydrophilic) group at one end, and a long nonpolar (hydrophobic) "tail." Their unique structural characteristics cause soap molecules to self-associate in aqueous solution and "entrap" or dissolve oil (dirt) molecules.*

Teacher's Notes

6. Tap water contains dissolved solids, in particular, magnesium and calcium salts. How do these salts change the surface tension of water?

   The surface tension of tap water is slightly greater than that of distilled or deionized water. Dissolved salts do not interfere with hydrogen bonding in water.

7. *(Optional)* Soaps and detergents work better in hot water than in cold water. What effect should increasing the temperature have on the surface tension of water? Explain in terms of the kinetic-molecular theory and the attractive forces between molecules.

   Soaps work better in hot water than in cold water because the surface tension of water decreases as the temperature increases. Increasing the temperature increases the average kinetic energy of the molecules. As the molecules move faster, they are more likely to break free from the uneven attractive forces at the surface of the liquid.

Teacher Notes

Vapor Pressure of Water
Effect of Temperature

Introduction

Dry air in buildings and homes causes many health problems during the winter months. Symptoms of low relative humidity include dry skin and rashes, sore throats, and coughs. The amount of water vapor that the air can hold depends on temperature. Why does warm air hold more water than cold air?

Concepts

- Evaporation and condensation
- Vapor pressure
- Kinetic-molecular theory
- Ideal gas law and Dalton's law

Background

The change of a substance from the liquid phase to the gaseous or vapor phase is called *evaporation*. The rate of evaporation of a liquid depends on the nature of the substance and varies with temperature. Imagine a liquid placed in a closed container at a given temperature. As molecules of liquid evaporate from the liquid surface, the number of molecules in the gas phase, and hence the pressure, will increase. As the number of gas molecules increases, some of the molecules may collide with the surface of the liquid and condense (reenter the liquid phase). Eventually, the rate of evaporation of the liquid molecules will become equal to the rate of condensation of the gas molecules, and the pressure of the vapor will reach a constant or equilibrium value. The *vapor pressure* of a liquid is defined as the pressure exerted by a vapor in equilibrium with the pure liquid at a specific temperature. Liquids with high vapor pressures are considered volatile—they will evaporate readily from an open container.

The kinetic-molecular theory provides the best model for explaining why different liquids have different vapor pressures. Molecules in the liquid state are in constant motion. They are close enough together, however, that attractive forces between molecules keep them in a definite volume and influence their properties. At a given temperature, some of the molecules will be moving fast enough and have sufficient kinetic energy to break free from the "ties that bind them" and escape into the vapor phase. If a liquid has strong forces of attraction between molecules, then only a few molecules will have enough energy to overcome the attractive forces and evaporate. Thus, liquids with strong attractive forces between molecules will have lower vapor pressures at any given temperature than liquids with weak intermolecular forces. The kinetic-molecular theory also explains why the vapor pressure of a liquid increases as the temperature increases. Increasing the temperature increases the average kinetic energy of the molecules. At higher temperatures, a larger number of molecules will be moving fast enough and have enough kinetic energy to overcome the forces of attraction in the liquid and enter the vapor phase. The number of molecules in the vapor phase above a liquid, and hence the vapor pressure, increases with increasing temperature.

This lab activity will require that your students have a good understanding of the ideal gas law and Dalton's law of partial pressures. These topics may have to be reviewed prior to starting this lab.

Vapor Pressure of Water – Page 2

Experiment Overview

The purpose of this experiment is to determine the vapor pressure of water at different temperatures. A small amount of air will be trapped in an inverted graduated cylinder and heated to about 80 °C in a hot water bath (Figure 1). As the temperature increases, the air in the graduated cylinder will expand and also become saturated with water vapor. The volume of gas will be measured at 5 °C-intervals as the water bath is cooled to about 50 °C. Finally, ice will be added to cool the trapped air to 0–5 °C and the volume of "dry air" will be recorded—at this temperature, the vapor pressure of water is so low that we can assume that the trapped gas contains only air (no water vapor).

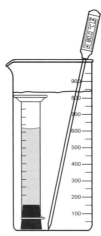

Figure 1.

Pre-Lab Questions

1. Is evaporation an exothermic or endothermic process? Explain in terms of the attractive forces between molecules.

2. The boiling point of a liquid is defined as the temperature at which the vapor pressure of the liquid is equal to the surrounding atmospheric pressure. How does the boiling point of water change as the air pressure decreases (for example, at higher altitudes)? Explain.

3. The vapor pressure of water at each temperature in this experiment will be calculated using the ideal gas law and Dalton's law of partial pressures.

    (a) Rearrange the ideal gas law equation to solve for the number of moles of "dry air" (n_{air}) in the graduated cylinder at 0–5 °C.

    (b) What is Dalton's law of partial pressures? Write an equation for Dalton's law that shows how the vapor pressure of water (P_{water}) can be determined if the partial pressure of air (P_{air}) and the total pressure (P_{atm}) are known. *Hint:* There are only two "gases" in the graduated cylinder (air and water vapor).

Materials

Barometer (optional)
Beaker, tall-form, 500-mL or 1-L
Beral-type pipet, jumbo
Gloves, heat-protective
Graduated cylinder, glass, 25-mL
Hot plate or Bunsen burner set-up

Ice
Ring stand and ring (optional)
Rubber stopper, 1-or 2-hole
Thermometer, digital
Stirring rod (plastic)
Water

Safety Precautions

Exercise care when working with the hot water bath. To avoid scalding yourself, wear heat-protective gloves or use a hot vessel gripping device when removing the beaker of hot water from the heat source. Turn the hot plate or burner off when not in use. Wear chemical splash goggles whenever working with chemicals, heat or glassware in the chemical laboratory.

Teacher Notes

Bring in a cake mix box or other bakery item box that provides instructions for baking at higher elevations. Ask your students why this is necessary.

Page 3 – Vapor Pressure of Water

Teacher Notes

Procedure

1. Fill a 500-mL or 1-L tall-form beaker about two-thirds full with *hot tap water* (45–50 °C).

2. Add 20–21 mL of distilled water to a 25-mL graduated cylinder and place a two-hole rubber stopper in the cylinder. Close the stopper holes with your finger and quickly invert and lower the graduated cylinder into the beaker filled with hot water. *Note:* There should be about 10 mL of air trapped in the inverted graduated cylinder.

3. Add more hot tap water as needed to the beaker so that the entire graduated cylinder is under water and the trapped air is also surrounded by the water bath (refer to Figure 1).

4. Place a digital thermometer in the hot water and heat the beaker on a hot plate at a medium setting or with a very low Bunsen burner flame until the water temperature is about 80 °C. *Note:* Do NOT allow the volume of gas to expand beyond the scale marked on the graduated cylinder.

5. Turn off the hot plate or Bunsen burner and carefully remove the heat source. Stir the water in the beaker to ensure an even distribution of heat. Record the temperature of the hot water bath and the volume of gas in the graduated cylinder in the data table.

6. Continue stirring the water and allow the beaker and contents to slowly cool. Measure and record the volume of gas and the temperature at approximately 5 °C-intervals until the temperature is about 50 °C. *Hint:* Don't cool too fast—the water and gas should reach thermal equilibrium.

7. Cold water or crushed ice may be added to speed up the cooling process. *It is important, however, to keep the water level in the beaker about the same throughout the experiment.* Remove some of the water as needed using a jumbo plastic pipet.

8. After the temperature has reached 50 °C, cool the water rapidly to about 0 °C by adding ice to the beaker. *Note:* Remove water as needed to keep the overall level of ice and water surrounding the graduated cylinder the same (see step 7).

9. When the temperature of the ice-water bath is between 0–5 °C, measure and record the volume of gas and the temperature in the data table.

10. Record the barometric pressure. Note the units!

Heating the water bath too fast (step 4) will cause the amount of gas in the cylinder to go off-scale. (The evaporation rate lags behind the heating rate.) If the volume does expand beyond the scale marked on the graduated cylinder, students must wait until it's back on scale to complete step 5.

Vapor Pressure of Water – Page 4

Name: _____

Class/Lab Period: _____

Vapor Pressure of Water

Data Table

Barometric Pressure	
Water Temperature	Volume of Gas

Post-Lab Calculations and Analysis

(Show all work on a separate sheet of paper and enter the results of the calculations in the Results Table.)

1. Gas law calculations require that the temperature be in Kelvin. Convert all of the temperature readings to *absolute temperature* units (Kelvin, K).

2. There is a small error in the measurement of the volume of gas caused by using the upside-down graduated cylinder (the meniscus is reversed). Correct all volume measurements by subtracting 0.2 mL from each volume reading. Convert this corrected volume measurement from milliliters (mL) to liters (L). Label this the *corrected volume* (V_{corr}).

3. Assume that "dry air" is the only gas in the inverted graduated cylinder at 0–5 °C—the vapor pressure of water is negligible at this temperature. Use the ideal gas law equation to calculate the *moles of air* (n_{air}) in the graduated cylinder at 0–5 °C. *Hint:* The pressure is equal to the barometric pressure.

4. The moles of air (n_{air}) in the graduated cylinder is constant throughout the experiment. Use the ideal gas law equation to calculate the *partial pressure of air* (P_{air}) in the graduated cylinder at each temperature between 50 and 80 °C. *Note:* Watch your units—be consistent!

5. Between 50 and 80 °C, there are two gases in the graduated cylinder—air and water vapor. The total pressure is equal to the barometric pressure (P_{atm}). Use Dalton's law to calculate the *vapor pressure of water* (P_{water}) at each temperature. (See *Pre-Lab Question* #3.) Convert the vapor pressure to mm Hg units.

Teacher Notes

6. Plot the vapor pressure of water versus temperature on the following graph. Draw the best *smooth-fit curve* through the data points. Notice that temperature is in Celsius!

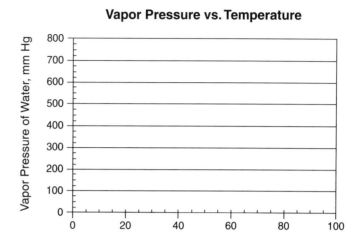

7. (a) The actual vapor pressure of water at 0 °C is 2 mm Hg. (b) The normal boiling point of a liquid is defined as the temperature at which the vapor pressure of the liquid is equal to 1 atm (760 mm Hg). Extend the curve shown above by drawing dotted lines to include the vapor pressure of water at 0 °C and 100 °C.

Results Table

Moles of "Dry Air" (n_{air})					
Temperature		V_{corr} (L)	P_{air} (atm)	P_{water} (atm)	P_{water} (mm Hg)
(°C)	(K)				

Teacher's Notes
Vapor Pressure of Water

Master Materials List *(for a class of 30 students working in pairs)*

- Barometer (optional) see *Lab Hints*
- Beakers, tall-form, 500-mL or 1-L, 15
- Beral-type pipets, jumbo, 15
- Gloves, heat-protective*
- Graduated cylinders, glass, 25-mL, 15
- Hot plates, 7*, or Bunsen burners
- Ice
- Ring stands and rings, 15 (optional)
- Rubber stoppers, 1-or 2-hole, 15
- Thermometers, digital, 15
- Stirring rods (plastic), 15
- Water

*Students may share gloves and hot plates.

Safety Precautions

Exercise care when working with the hot water bath. To avoid scalding, wear heat-protective gloves or use a hot-vessel gripping device when removing the beaker of hot water from the heat source. Turn the hot plate or burner off when not in use. Wear chemical splash goggles whenever working with chemicals, heat or glassware in the chemical laboratory.

Disposal

No special disposal procedures are required.

Lab Hints

- The laboratory work for this experiment may require a double 50-minute class period. Heating the water is a time consuming process, and the water bath should be cooled slowly in order to achieve good heat transfer. Having hot water ready at the start of the laboratory will save time.

- The experiment may also be done in 10-mL graduated cylinders, although we found that the vapor pressure results were more accurate when 25-mL graduated cylinders were used.

- It is not practical to raise the graduated cylinder out of the water bath to equalize the pressure of the gas inside the cylinder with atmospheric pressure. The temperature of the gas cools rapidly when the graduated cylinder is lifted out of the bath, giving rise to large errors in the resulting volume measurements. The height difference between the gas level in the graduated cylinder and the water level in the beaker is thus a major source of error in this experiment. The pressure inside the cylinder is slightly greater than atmospheric pressure because the trapped gas is supporting the extra height of the column of water (assuming the water level in the beaker is above the gas level in the graduated cylinder). The pressure differential may be accounted for, if desired, by measuring the height (h) in millimeters of the water level above the gas level in the graduated cylinder and dividing by 13.6. The results in the *Sample Data* section were obtained without applying this pressure correction.

$$P_{cylinder} = P_{atm} + h \text{ (mm H}_2\text{O)} \times (1 \text{ mm Hg}/13.6 \text{ mm H}_2\text{O})$$

(P is in units of mm Hg)

Do not use economy-choice graduated cylinders—the scale is not precise enough for this lab.

Teacher's Notes

Teacher Notes

Teaching Tip

- The effect of temperature on the vapor pressure of water is important in the study of weather (meteorology). The relative humidity is defined as the percentage of moisture that the air can hold compared to the maximum that it can hold at a specific temperature (based on the phase diagram for the saturation vapor pressure of water as a function of temperature). The dew point is the temperature at which the present amount of humidity in the air would reach 100% relative humidity and thus start to condense.

Answers to Pre-Lab Questions *(Student answers will vary.)*

1. Is evaporation an exothermic or endothermic process? Explain in terms of the attractive forces between molecules.

    Evaporation is an endothermic process—heat energy is required in order to break the attractive forces between the molecules in the liquid state and allow them to "escape" into the vapor phase.

2. The boiling point of a liquid is defined as the temperature at which the vapor pressure of the liquid is equal to the surrounding atmospheric pressure. How does the boiling point of water change as the air pressure decreases (for example, at higher altitudes)? Explain.

    The boiling point of water is lower than 100 °C at higher altitudes (as the air pressure decreases). **Note to teachers:** *At the summit of Mt. Everest, for example, the atmospheric pressure is only 240 mm Hg, and water boils at 71 °C. (That's the temperature at which the vapor pressure of water equals 240 mm Hg.) Vapor pressure increases with increasing temperature.*

3. The vapor pressure of water at each temperature in this experiment will be calculated using the ideal gas law and Dalton's law of partial pressures.

    (a) Rearrange the ideal gas law equation to solve for the number of moles of "dry air" (n_{air}) in the graduated cylinder at 0–5 °C.

    $$n_{air} = \frac{P_{atm} \times V}{RT}$$

    (b) What is Dalton's law of partial pressures? Write an equation for Dalton's law that shows how the vapor pressure of water (P_{water}) can be determined if the partial pressure of air (P_{air}) and the total pressure (P_{atm}) are known. *Hint:* There are two gases in the graduated cylinder (air and water vapor).

    According to Dalton's law, the total pressure of a mixture of gases is equal to the sum of the partial pressures of the components in the gas mixtures.

    $$P_{atm} = P_{air} + P_{water}$$

    $$P_{water} = P_{atm} - P_{air}$$

Teacher's Notes

Sample Data

Student data will vary.

Data Table

Barometric Pressure	746 mm Hg
Water Temperature	**Volume of Gas**
80.8 °C	18.4 mL
74.8 °C	15.2 mL
69.9 °C	13.2 mL
64.6 °C	12.3 mL
59.8 °C	11.2 mL
54.7 °C	10.6 mL
49.5 °C	10.2 mL
5.0 °C	7.8 mL

Answers to Post-Lab Calculations and Analysis *(Student answers will vary.)*

1. Gas law calculations require that the temperature be in Kelvin. Convert all of the temperature readings to *absolute temperature* units (Kelvin, K).

   K = °C + 273.2
   Sample calculation: 5.0 °C + 273.2 = 278.2 K
   See the Sample Results Table for the results of the calculations.

2. There is a small error in the measurement of the volume of gas caused by using the upside-down graduated cylinder (the meniscus is reversed). Correct all volume measurements by subtracting 0.2 mL from each volume reading. Convert this corrected volume measurement from milliliters (mL) to liters (L). Label this the *corrected volume* (V_{corr}).

   V_{corr} = V − 0.2 mL
   Sample calculation: V_{corr} = 7.8 mL − 0.2 mL = 7.6 mL (0.0076 L) at 5.0 °C (278 K)

   See the Sample Results Table for the results of the calculations.

3. Assume that "dry air" is the only gas in the inverted graduated cylinder at 0–5 °C—the vapor pressure of water is negligible at this temperature. Use the ideal gas law equation to calculate the *moles of air* (n_{air}) in the graduated cylinder at 0–5 °C. *Hint:* The pressure is equal to the barometric pressure.

   $$n_{air} = \frac{P_{atm} \times V}{RT} = \frac{(0.982 \text{ atm})(0.0076 \text{ L})}{(0.0821 \text{ L atm/mole K})(278 \text{ K})} = 3.27 \times 10^{-4} \text{ moles}$$

Teacher's Notes

4. The moles of air (n_{air}) in the graduated cylinder is constant throughout the experiment. Use the ideal gas law equation to calculate the *partial pressure of air* (P_{air}) in the graduated cylinder at each temperature between 50 and 80 °C. *Note:* Watch your units—be consistent!

   Sample calculation for T = 80.8 °C (354 K), V_{corr} = 18.2 mL (0.0182 L):

   $$P_{air} = \frac{(3.27 \times 10^{-4} \text{ moles})(0.0821 \text{ L atm/mole K})(354 \text{ K})}{(0.0182 \text{ L})} = 0.522 \text{ atm}$$

   See the Sample Results Table for the results of the other calculations.

5. Between 50 and 80 °C, there are two gases in the graduated cylinder—air and water vapor. The total pressure is equal to the barometric pressure (P_{atm}). Use Dalton's law to calculate the *vapor pressure of water* (P_{water}) at each temperature. (See *PreLab Question #3.*) Convert the vapor pressure to mm Hg units.

   Sample calculation for T = 80.8 °C (354 K), P_{atm} = 0.982 atm, P_{air} = 0.522 atm:

   $$P_{atm} = P_{air} + P_{water}$$

   $$P_{water} = 0.982 \text{ atm} - 0.522 \text{ atm} = 0.46 \text{ atm}$$

   $$P_{water} = 0.46 \text{ atm} \times \frac{760 \text{ mm Hg}}{1 \text{ atm}} = 350 \text{ mm Hg}$$

   See the Sample Results Table for the results of the other calculations.

6. Plot the vapor pressure of water versus temperature on the following graph. Draw the best *smooth-fit curve* through the data points. Notice that temperature is in Celsius!

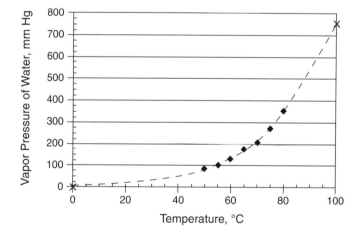

Vapor Pressure of Water

Teacher's Notes

7. (a) The actual vapor pressure of water at 0 °C is 2 mm Hg. (b) The normal boiling point of a liquid is defined as the temperature at which the vapor pressure of the liquid is equal to 1 atm (760 mm Hg). Extend the curve shown above by drawing dotted lines to include the vapor pressure of water at 0 °C and 100 °C.

See the graph.

Sample Results Table

Moles of "Dry Air"					3.27×10^{-4} moles
Temperature		V_{corr} (L)	P_{air} (atm)	P_{water} (atm)	P_{water} (mm Hg)
(°C)	(K)				
80.8 °C	354 K	0.0182 L	0.521 atm	0.461 atm	350 mm Hg
74.8 °C	348 K	0.0150 L	0.622 atm	0.360 atm	274 mm Hg
69.9 °C	343.1 K	0.0130 L	0.707 atm	0.275 atm	209 mm Hg
64.6 °C	337.8 K	0.0121 L	0.748 atm	0.234 atm	178 mm Hg
59.8 °C	333 K	0.0110 L	0.811 atm	0.171 atm	130 mm Hg
54.7 °C	327.9 K	0.0194 L	0.845 atm	0.137 atm	104 mm Hg
49.5 °C	322.7 K	0.0100 L	0.865 atm	0.117 atm	89 mm Hg
5.0 °C	278.2 K	0.0076 L	0.981 atm	NA	NA

Teacher Notes

How Cool Is That?
Evaporation of Liquids

Introduction

It's a hot and sunny summer day, and you step out of the pool, cool and refreshed. Soon, however, your teeth start chattering and your lips turn blue. Water evaporating from your skin draws heat from your body, leaving you feeling cold. The "cooling effect of evaporation" is nature's most important way of cooling not only our bodies but also the Earth! How cool is evaporation?

Concepts

- Evaporation
- Polar vs. nonpolar compounds
- Kinetic-molecular theory
- Hydrogen bonding

Background

Vaporization is the process by which a substance changes from a liquid to a gas or vapor. When vaporization occurs gradually from the surface of a liquid, it is called *evaporation*. Evaporation is an endothermic process—energy is required for molecules to leave the liquid phase and enter the gas phase. The most common way to provide energy for the vaporization of a liquid is by heating it. Water evaporating from the Earth's oceans, for example, absorbs heat energy from the sun and helps to moderate the temperature around large bodies of water. When the heat energy for vaporization comes from the surroundings rather than from external heating, the temperature of the surroundings will decrease when a liquid evaporates. This is the origin of the cooling effect of evaporation. Water evaporating from the surface of the skin by perspiration, for example, cools the body.

Evaporation and the cooling effect of evaporation may be explained using the *kinetic-molecular theory*. According to this model, molecules in the liquid state are in constant motion, and interactions among neighboring molecules influence the motion of the molecules and the properties of the liquid. The temperature of a substance is proportional to the *average kinetic energy,* and thus the average speed, of the molecules. Evaporation occurs when fast-moving molecules near the surface of a liquid have enough energy to break free of their interactions with neighboring molecules and "escape" into the gas phase. Molecules with the highest kinetic energy evaporate and become gas molecules. The average kinetic energy, and thus the temperature, of the remaining molecules decreases—a liquid cools as it evaporates. This phenomenon is known as "evaporative cooling." The rate of evaporation of a liquid increases at higher temperatures, because more molecules have enough energy to break free of the liquid's surface.

The rate of evaporation of a liquid depends on the nature of the liquid and the type of attractive forces between molecules. Strong intermolecular attractions hold the molecules in a liquid more tightly. Liquids with weak intermolecular attractive forces have low heats of vaporization and are volatile—they evaporate easily. Liquids with strong intermolecular attractive forces evaporate more slowly, because a greater amount of energy is needed to overcome the attractive forces between the molecules.

How Cool Is That? – Page 2

Nonpolar compounds generally have very weak attractive forces, called London dispersion forces, between molecules. The strength of London dispersion forces increases in a regular manner as the size of the molecules increase. Dipole interactions occur when polar molecules are attracted to one another. Because dipole interactions are stronger than dispersion forces, polar compounds generally have higher heats of vaporization and evaporate more slowly than nonpolar compounds (assuming that the molecules have similar molar masses). Hydrogen bonding represents a special case of dipole interactions, in which F–H, O–H and N–H groups in molecules associate very strongly with electronegative atoms in adjacent molecules. Hydrogen bonds are the strongest type of intermolecular attractive forces. Hydrogen bonding in water, for example, leads to a high degree of association among water molecules in the liquid and solid state. As a result, water is a very unusual liquid in many ways. It has an unusually high heat of vaporization and a very high boiling point compared to other compounds that are about the same size or have similar structures. Evaporation of water acts as a "heat sink" for energy from the sun. A significant portion of the sun's energy that reaches Earth is spent evaporating water from the oceans, lakes, and rivers rather than warming the Earth.

Experiment Overview

The purpose of this experiment is to measure the temperature changes that occur when different liquids evaporate and to compare their rates of evaporation. The experiment will be carried out using a computer interface with temperature probes or sensors that have been soaked in various liquids. The temperature will be recorded versus time as the liquids evaporate. Liquids will be compared pair-wise (e.g., polar versus nonpolar, presence or absence of hydrogen bonding, etc.) and the results will be analyzed in terms of the strength of attractive forces.

Pre-Lab Questions

Complete the following table summarizing the properties of the liquids to be studied.

Compound	Structural Formula	Molar Mass	Polar vs. Nonpolar	Hydrogen Bonding Ability
Acetone				
Ethyl Alcohol				
Heptane				
Hexane				
Isopropyl Alcohol				
Methyl Alcohol				

Teacher Notes

In comparing the strength of attractive forces between different types of molecules, it is very important to keep the size of the molecules similar. Dispersion forces increase for all types of molecules as they increase in size.

Page 3 – **How Cool Is That?**

Teacher Notes

Materials

Acetone, $(CH_3)_2CO$, 2 mL
Ethyl alcohol, CH_3CH_2OH, 2 mL
Heptane, C_7H_{16}, 2 mL
Hexane, C_6H_{14}, 2 mL
Isopropyl alcohol, $(CH_3)_2CHOH$, 2 mL
Methyl alcohol, CH_3OH, 2 mL
Computer interface system (LabPro™)
Data collection software (LoggerPro™)

Pipets, 6
Corks or stoppers to fit test tubes, 6
Filter paper or cotton gauze, 11-cm, 2
Rubber bands, small (orthodontic-type), 6
Scissors
Temperature probes or sensors, 2
Test tubes, small, 6
Test tube rack

Safety Precautions

Acetone, methyl alcohol, ethyl alcohol, hexane, heptane, and isopropyl alcohol are flammable liquids and a dangerous fire risk. Avoid contact of all liquids with heat, flames or other sources of ignition. Methyl alcohol is toxic by ingestion. Acetone, ethyl alcohol, and heptane are slightly toxic by ingestion or inhalation. Addition of denaturant makes ethyl alcohol poisonous—it cannot be made non-poisonous. Do not allow chemicals to come into contact with eyes and skin. Perform this experiment in a well-ventilated lab only. Wear chemical splash goggles, chemical-resistant gloves, and a chemical-resistant apron. Wash hands thoroughly with soap and water before leaving the laboratory.

Procedure

1. Cut six small pieces of filter paper or cotton gauze, approximately 3 cm square each.

2. Label six test tubes #1–6 and place them in a test tube rack. Obtain 2–3 mL of the appropriate solvent in each test tube, according to the following scheme. Stopper the test tubes with corks or rubber stoppers until needed.

Test tube	1	2	3	4	5	6
Solvent	Hexane	Heptane	Methyl alcohol	Ethyl alcohol	Acetone	Isopropyl alcohol

3. Wrap one filter paper square around each temperature probe and secure the filter paper with a small rubber band.

4. Plug the temperature probes (sensors) into CH1 and CH2 on the interface system (LabPro) and connect the LabPro to the computer or calculator.

5. Open the data collection software (LoggerPro) on the computer. Select *Setup* and *Sensors* from the main screen and choose the appropriate temperature probe for both CH1 and CH2.

6. Select *Experiment* and *Data Collection* from the main screen and choose *Mode—Time Based*.

7. Enter the following choices in the dialog box to set the conditions for the experiment: *Length* – 180 sec; *Sampling speed* – 2 seconds/sample.

8. Choose *Save* from the *File* menu and save the experiment file. Make sure *Over-sampling* is not checked.

This experiment may also be done using digital thermometers and a watch or timer with a second hand. Try to measure the temperature every 10–15 seconds, at least.

How Cool Is That? – Page 4

Teacher Notes

9. Select *Experiment* and *Remote* from the main screen. Check the *OK* box for remote setup. (The yellow light on the LabPro will come on at this point, indicating that the LabPro is ready.)

10. Disconnect the LabPro from the computer.

11. Place the CH1 temperature probe into test tube #1 (hexane) and the CH2 temperature probe into test tube #2 (heptane). The liquid level in each test tubes should be above the filter paper to ensure that the paper is thoroughly soaked with liquid. Allow the temperature probes to soak in the liquid for about 30 seconds. (*This is Trial A.*)

12. Press the *Start/Stop* button on the LabPro.

13. The yellow light will go off and a green light will blink as each data point is collected. Collect 3–4 data points, then remove the temperature probes from the test tubes and carefully extend the probes over the test tube rack as shown in Figure 1. Avoid jostling the temperature probes (air drafts may change the rate at which the liquids evaporate).

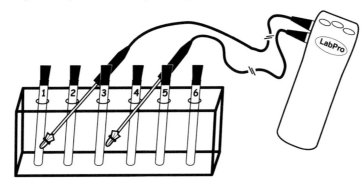

Figure 1.

14. The yellow light on the LabPro will briefly flash when the sampling period is over and data collection is done.

15. Open the experiment file on the computer and select *Remote* followed by *Retrieve Data*. Reconnect the LabPro to the computer and hit *OK*.

16. The data will be displayed on the computer screen in two formats—a table of Time versus Temperature and a graph of Temperature versus Time. Save the experiment file and print the table and graph, if possible.

17. Remove the filter paper squares from the temperature probes and carefully dry the probes with a paper towel.

18. Repeat steps 3–17 using methyl alcohol (test tube #3) and ethyl alcohol (test tube #4) in CH1 and CH2, respectively. *(This is Trial B.)*

19. Repeat steps 3–17 using acetone (test tube #5) and isopropyl alcohol (test tube #6) in CH1 and CH2, respectively. *(This is Trial C.)*

Teacher Notes

Name: _____

Class/Lab Period: _____

How Cool Is That?

Data and Results Table

	Evaporation of Liquids	Trial A	Trial B	Trial C
CH1	Liquid			
	Initial temperature			
	Minimum temperature			
	Temperature change due to evaporation			
CH2	Liquid			
	Initial temperature			
	Minimum temperature			
	Temperature change due to evaporation			

Post-Lab Questions

1. Describe in words a typical temperature versus time graph for the evaporation of a liquid—be as specific as possible. Explain the graph in terms of the "cooling effect of evaporation."

2. Summarize the results of this experiment in the Data and Results Table. Subtract the initial temperature from the minimum temperature to determine the temperature change due to evaporation. What is the relationship between the temperature change due to evaporation and the rate of evaporation of a liquid? Explain.

How Cool Is That? – Page 6

3. Compare the results obtained for hexane and heptane in Trial A. Which compound evaporated more quickly? How are these compounds similar? How are they different? Which compound has stronger attractive forces? Explain.

4. Compare the results obtained for methyl alcohol and ethyl alcohol in Trial B. Which compound evaporated more quickly? How are these compounds similar? How are they different? Which compound has stronger attractive forces? Explain.

5. Compare the results obtained for acetone and isopropyl alcohol in Trial C. Which compound evaporated more quickly? How are these compounds similar? How are they different? Which compound has stronger attractive forces? Explain.

6. (a) Is there a pattern between the molar mass of a compound and the temperature change observed due to evaporation? (b) Why would it not be fair to conclude that "dispersion forces are stronger than hydrogen bonding" by comparing the results for hexane and methyl alcohol in this experiment?

7. Rank the six liquids tested from most volatile to least volatile based on their observed temperature changes due to evaporation.

8. *(Optional)* Look up the boiling points of the six compounds tested in this experiment. Is there a relationship between the rate of evaporation of a liquid and its boiling point?

Teacher Notes

Teacher's Notes
How Cool Is That?

Master Materials List *(for a class of 30 students working in groups of three)**

Acetone, $(CH_3)_2CO$, 25 mL	Pipets, 60
Ethyl alcohol, CH_3CH_2OH, 25 mL	Corks or stoppers to fit test tubes, 60
Heptane, C_7H_{16}, 25 mL	Filter paper or cotton gauze, 11-cm, 20
Hexane, C_6H_{14}, 25 mL	Rubber bands, small (orthodontic-type), 60
Isopropyl alcohol, $(CH_3)_2CHOH$, 25 mL	Scissors, 10
Methyl alcohol, CH_3OH, 25 mL	Temperature probes or sensors, 20
Computer interface systems (LabPro™), 10	Test tubes, small, 60
Data collection software (LoggerPro™)	Test tube racks, 10

*Adjust the working group size to accommodate the number of temperature probes or sensors and the number of LabPro interfaces available for the classroom. See the Lab Hints section.

Safety Precautions

Acetone, methyl alcohol, ethyl alcohol, hexane, heptane, and isopropyl alcohol are flammable liquids and a dangerous fire risk. Avoid contact of all liquids with heat, flames or other sources of ignition. Methyl alcohol is toxic by ingestion. Acetone, ethyl alcohol, and heptane are slightly toxic by ingestion or inhalation. Addition of denaturant makes ethyl alcohol poisonous—it cannot be made non-poisonous. Do not allow chemicals to come into contact with eyes and skin. Perform this experiment in a well-ventilated lab only. Wear chemical splash goggles, chemical-resistant gloves, and a chemical-resistant apron. Please review current Material Safety Data Sheets for additional safety, handling, and disposal information. Remind students to wash hands thoroughly with soap and water before leaving the laboratory.

Disposal

Please consult your current *Flinn Scientific Catalog/Reference Manual* for general guidelines and specific procedures governing the disposal of laboratory waste. Isopropyl alcohol solutions may be rinsed down the drain with plenty of water according to Flinn Suggested Disposal Method #26b.

Lab Hints

- The laboratory work for this experiment can probably be completed within 20–30 minutes. Each trial takes only a few minutes to set up and collect data. Adjust the working group size as needed to accommodate the number of technology stations or temperature probes available in the lab. Alternatively, the experiment may be paired during the same period with a complementary experiment, such as "Properties of Liquids," which is also fairly short. Stagger the labs so that one half of the class works on one experiment, the other half on the second experiment, and then switch.

- If each group takes a plastic or glass pipet of solution back to their work area, about 60 pipets will be needed. If each group brings their well plate to the solution bottles, then only 6 or 12 pipets will be needed. Rubberband test tubes to reagent bottles so students can place used pipets in an uncontaminated storate site.

When using organic solvents, glass Pasteur pipets or medicine droppers work better than plastic Beral-type pipets.

Teacher's Notes

- "Hexanes," a mixture of *n*-hexane and other isomers, may be used as the source of hexane for Part A. The boiling point of the mixture (68–70 °C) is close enough to that of *n*-hexane (69 °C) that its use should not be a problem. In general, branched-chain alkanes have lower boiling points than their straight-chain isomers—the boiling points of *n*-hexane, 3-methylpentane (1 branch) and 2,3-dimethylbutane (2 branches) are 69 °C, 63 °C, and 58 °C, respectively. (This is attributed to the lower surface area of spherical, branched-chain compounds compared to their rod-like, straight-chain isomers.)

- Isopropyl alcohol is recommended rather than *n*-propyl alcohol (1-propanol) because it is more readily available in high school chemistry labs and because it is less toxic by inhalation (the TLV is 983 mg/m^3 for isopropyl alcohol compared to 492 mg/m^3 for *n*-propyl alcohol).

- Temperature measurements may be made using bare temperature probes that have been dipped into the liquid. The liquid will evaporate very quickly, however, so the temperature will bottom out after about 30 seconds and will then start to increase.

- Many different solvent pairs may be tested in this experiment. We tried to balance availability and relative toxicity of different solvents against the desire for good pair-wise comparisons (e.g., effect of molar mass in alkanes, polar versus nonpolar, presence or absence of hydrogen bonding). Mindful of the inhalation hazard for both students and the teacher in this "evaporation lab," we decided to limit the total number of solvents to no more than six (three trials). Hexane was selected rather than pentane for the alkane series because pentane has a very low flash point and is narcotic in high concentrations. *n*-Butyl alcohol was not included in the alcohol series because of its strong odor and low TLV. Teachers who want to further limit the number of solvents used in the lab may want to restrict the solvents to only one category or variable, such as the series of methyl, ethyl, propyl, and butyl alcohol.

Teaching Tip

- Evaporative cooling is one of the oldest and most energy efficient methods of cooling a home. A typical evaporative cooler uses only about one-fourth as much electricity as an air conditioner during peak demand. It also does not require ozone-depleting gases, and is thus an environmentally friendly alternative to air conditioning. An evaporative cooler basically consists of a large fan that draws outside air through a water-soaked pad—evaporation of water cools the air and increases the moisture content in the air. The resulting temperature decrease that takes place depends on the temperature and the relative humidity of the incoming air. Outside air at 100 °F and a relative humidity of 30%, for example, will be cooled to about 82 °F. Direct evaporative coolers are a viable alternative to air conditioners in hot and dry climates such as the desert Southwest. They are not useful in hot and humid areas such as the Midwest or the South Atlantic region.

Teacher's Notes

Teacher Notes

Answers to Pre-Lab Questions

Compound	Structural Formula	Molar Mass	Polar vs. Nonpolar	Hydrogen Bonding Ability
Acetone	$CH_3-CO-CH_3$	58 g/mole	Polar	No
Ethyl Alcohol	CH_3-CH_2-OH	46 g/mole	Polar	Yes
Heptane	$CH_3-CH_2-CH_2-CH_2-CH_2-CH_2-CH_3$	100 g/mole	Nonpolar	No
Hexane	$CH_3-CH_2-CH_2-CH_2-CH_2-CH_3$	86 g/mole	Nonpolar	No
Isopropyl Alcohol	$CH_3-CHOH-CH_3$	60 g/mole	Polar	Yes
Methyl Alcohol	CH_3-OH	32 g/mole	Polar	Yes

*Acetone molecules do not form hydrogen bonds with other acetone molecules.
Acetone molecules may form hydrogen bonds with O—H group donor molecules.

How Cool Is That?

Teacher's Notes

Sample Data

Student data will vary.

Data and Results Table

	Evaporation of Liquids	Trial A	Trial B	Trial C
CH1	Liquid	Hexane	Methyl alcohol	Acetone
	Initial temperature	22.8 °C	21.6 °C	21.8 °C
	Minimum temperature	7.0 °C	5.1 °C	4.5 °C
	Temperature change due to evaporation	−15.8 °C	−16.5 °C	−17.3 °C
CH2	Liquid	Heptane	Ethyl alcohol	Isopropyl alcohol
	Initial temperature	23.0 °C	22.6 °C	22.5 °C
	Minimum temperature	15.3 °C	12.6 °C	14.3 °C
	Temperature change due to evaporation	−7.7 °C	−10.0 °C	−8.2 °C

Sample Graphs

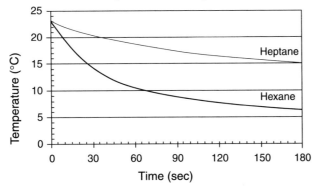

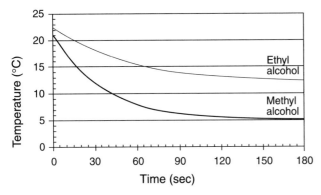

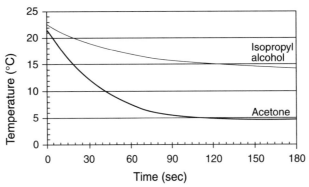

Teacher's Notes

Teacher Notes

Answers to Post-Lab Questions (*Student answers will vary.*)

1. Describe in words a typical temperature versus time graph for the evaporation of a liquid—be as specific as possible. Explain the graph in terms of the "cooling effect of evaporation."

   The initial temperature of the liquid is constant at or near room temperature for the first 10 seconds or so, corresponding to the first 3–4 data points, when the temperature probe is still in the liquid (see step 13 in the Procedure). There is then a sharp decrease in the temperature over the next 30 seconds as the liquid begins to evaporate. The temperature will usually continue to decrease over the entire time period (180 sec), although the rate at which the temperature decreases begins to slow down or level off after about 100 seconds. The minimum temperature is generally observed after about 150 seconds. The temperature decrease occurs because a liquid cools as it evaporates and absorbs heat from the surroundings. The rate of evaporation of a liquid decreases at lower temperatures. The rate at which the temperature decreases thus reflects the rate of evaporation—as the liquid cools, it evaporates more slowly.

2. Summarize the results of this experiment in the Data and Results Table. Subtract the initial temperature from the minimum temperature to determine the temperature change due to evaporation. What is the relationship between the temperature change due to evaporation and the rate of evaporation of a liquid? Explain.

   See the Data and Results Table. The observed temperature change due to evaporation correlates with the ease of evaporation of a liquid and its rate of evaporation. Liquids that evaporate very fast show large temperature changes (lower final temperatures).

3. Compare the results obtained for hexane and heptane in Trial A. Which compound evaporated more quickly? How are these compounds similar? How are they different? Which compound has stronger attractive forces? Explain.

   Hexane reached a lower minimum temperature than heptane. This means that hexane evaporated more quickly than heptane. Hexane and heptane have very similar structures—they are both nonpolar hydrocarbons consisting of C—C and C—H bonds. Heptane has one more –CH$_2$– group in its "chain" and thus has stronger attractive forces than hexane. (Dispersion forces are stronger for "bigger" molecules.)

4. Compare the results obtained for methyl alcohol and ethyl alcohol in Trial B. Which compound evaporated more quickly? How are these compounds similar? How are they different? Which compound has stronger attractive forces? Explain.

   Methyl alcohol reached a lower minimum temperature than ethyl alcohol. This means that methyl alcohol evaporated more quickly than ethyl alcohol. Methyl and ethyl alcohol have similar structures—they both contain –OH groups capable of hydrogen bonding. Ethyl alcohol has one more –CH$_2$– group in its structure and thus has stronger attractive forces than methyl alcohol. (All other things being equal, attractive forces are stronger for "bigger" molecules.)

5. Compare the results obtained for acetone and isopropyl alcohol in Trial C. Which compound evaporated more quickly? How are these compounds similar? How are they different? Which compound has stronger attractive forces? Explain.

   Acetone reached a lower minimum temperature than isopropyl alcohol. This means that acetone evaporated more quickly than isopropyl alcohol. Acetone and isopropyl alcohol have similar molar masses and both are polar compounds. Isopropyl alcohol, however, has an –OH group in its structure and is thus capable of forming hydrogen bonds with neighboring molecules. Isopropyl alcohol has stronger attractive forces than acetone.

6. (a) Is there a pattern between the molar mass of a compound and the temperature change observed due to evaporation? (b) Why would it not be fair to conclude that "dispersion forces are stronger than hydrogen bonding" by comparing the results for hexane and methyl alcohol in this experiment?

   (a) In general, compounds with larger molar masses exhibited smaller temperature changes due to evaporation (they evaporated more slowly). This was true regardless of which series was being compared—hexane and heptane in Trial A or methyl alcohol and ethyl alcohol in Trial B.

   (b) Hexane and methyl alcohol not only have different structures, they also have different molar masses. Hexane exhibited a smaller temperature change than methyl alcohol. The overall strength of the attractive forces is thus slightly greater for hexane than for methyl alcohol, but this is because hexane is a much bigger molecule than methyl alcohol, not because dispersion forces are stronger than hydrogen bonds. The molar mass of hexane is 86 g/mole, compared to 32 g/mole for methyl alcohol. A better comparison would be hexane and isopropyl alcohol. Hexane has a larger molar mass than isopropyl alcohol (86 versus 60 g/moles), but it evaporated much more quickly, that is, it has weaker attractive forces.

7. Rank the six liquids tested from most volatile to least volatile based on their observed temperature changes due to evaporation.

   From most volatile to least volatile:

   Acetone > methyl alcohol > hexane > ethyl alcohol > isopropyl alcohol > heptane

8. *(Optional)* Look up the boiling points of the six compounds tested in this experiment. Is there a relationship between the rate of evaporation of a liquid and its boiling point?

   The rate of evaporation increases as the boiling point decreases. Thus, compounds that have lower boiling points exhibited larger temperature changes due to evaporation. **Note to teachers:** *The data are tabulated on the following page. There is a relationship between the temperature change due to evaporation and the boiling point, but it is not linear.*

Teacher's Notes

Teacher Notes

Solvent	Temperature Change Due to Evaporation	Boiling Point
Acetone	−17.3 °C	56.5 °C
Methyl Alcohol	−16.5 °C	64.7 °C
Hexane	−15.8 °C	69.0 °C
Ethyl Alcohol	−10.0 °C	78.5 °C
Isopropyl Alcohol	−8.2 °C	82.5 °C
Heptane	−7.7 °C	98.4 °C

Teacher's Notes

Teacher Notes

Teacher Notes

Teaching with Toys
Part A. Drinking Bird

Introduction

Homer Simpson has called the "drinking bird" toy the greatest invention in the world! Let your students discover evaporation and the effect of temperature on vapor pressure by "playing" with this popular toy in the classroom or lab.

Concepts

- Evaporation
- Vapor pressure

Materials

Cup or beaker
Drinking bird
Vacuum or bell jar (optional)
Water

Safety Precautions

The drinking bird is fragile and will break if dropped. The liquid inside the bird is typically a volatile hydrocarbon. Although the exact liquid that is used may change frequently depending on the manufacturer's formulation, always assume that the liquid is flammable and toxic by ingestion. Wear safety goggles when handling the drinking bird.

Procedure

1. Assemble the drinking bird by placing the metal prongs into the tops of the legs. Be sure that the beak is facing the same way as the feet.

2. The metal pivot should be approximately halfway between the bottom bulb and the red-colored neck of the drinking bird. The metal pivot should also be bent slightly so that the bird tends to lean forward. This will keep the drinking bird from doing back flips (and breaking!).

3. Briefly submerge the entire head of the drinking bird into a cup of tap water.

4. Set the bird next to the cup, with the beak facing the cup, so that the bird will appear to be drinking from the cup when it tips over.

5. Observe—the liquid inside the drinking bird will slowly rise up the center tube until the bird becomes top-heavy. This will cause the bird to tip over as if to drink the water from the cup. When the bird is in the horizontal position, air bubbles will enter the tube and the liquid will drain back into the bottom bulb. The bottom-heavy bird will then right itself, and the entire process will begin again.

Disposal

None required—save all materials for future use.

Teaching with Toys – Page 2

Teacher Notes

Tip

- Ask students to suggest and test conditions which will cause the bird to stop drinking. There are several scenarios: (1) Remove most of the water from the cup. The bird will stop drinking when there is not enough water left in the cup to wet the bird's head after it tips over. (2) Place a large bell jar over the bird and the cup. The bird will stop drinking when the relative humidity approaches saturation and the water will no longer evaporate from the bird's head.

Discussion

There are three main principles at work (or in play!) with the drinking bird toy.

- Evaporation is an endothermic process—when a liquid evaporates, it absorbs heat from its surroundings. This gives rise to the cooling effect of evaporation.
- The vapor pressure of a liquid decreases as the temperature decreases.
- Differences in gas pressure will cause a liquid to rise or fall in a tube that is closed at one end.

The action of the drinking bird toy is summarized in the following series of steps.

1. The bird's head is dipped into water.

2. As the bird is standing upright, water begins to evaporate from the bird's beak. The temperature of the "upper bulb" (the bird's head) decreases due to the cooling effect of evaporation.

3. The pressure of the vapor in the upper bulb decreases as the temperature is reduced.

4. The vapor pressure is now greater in the lower bulb (which has not cooled) than in the upper bulb. The pressure difference forces liquid up the center tube. See Figure 1.

5. The vapor pressure of the liquid in the lower bulb remains constant (there is no temperature change). The pressure in the upper bulb continues to decrease due to "evaporative cooling." The liquid level in the center tube continues to rise until . . .

6. The drinking bird becomes top-heavy, causing it to tip over and begin drinking from the cup again.

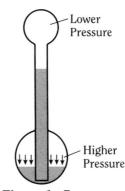

Figure 1. Pressure difference forces liquid up the tube.

7. When the bird is in the horizontal position, gas bubbles enter into the tube. This equalizes the gas pressure in the upper and lower bulbs, and the liquid in the upper bulb drains back into the lower bulb due to gravity. (See Figure 2.)

8. When the bird once again becomes bottom-heavy, it rights itself, and the process begins again.

Figure 2. Gas bubbles enter the tube to equalize the pressure.

Flinn ChemTopic™ Labs — Solids and Liquids

Teacher Notes

Part B. Hand Boiler

Introduction

A hand boiler is a closed glass container consisting of two bulbs that are connected by spiral glass tubing and partially filled with a colored liquid. In normal use, the liquid inside the hand boiler will boil with just a touch of the hand. A little ingenuity, however, converts the hand boiler into a distillation apparatus that can be used to separate the volatile organic liquid from the colored dye.

Concepts

- Vapor pressure
- Distillation

Materials

Beaker or cup
"Hand boiler" toy
Ice water

Safety Precautions

The hand boiler is fragile and will break if dropped. The liquid inside the hand boiler is a volatile organic solvent. Although the exact liquid that is used may change depending on the manufacturer's formulation, always assume that the liquid is flammable and toxic by ingestion. Wear safety goggles when handling the hand boiler.

Procedure

1. Transfer all of the red liquid into the large (lower) bulb. This may take some effort due to the bends and loops in the glass tubing, but the demonstration works best when all of the red liquid is in the large bulb.

2. Tilt the hand boiler to turn it upside down as shown in Figure 1. The liquid will not drain into the smaller bulb because the central glass tubing extends up into the large bulb.

3. Cup the hand boiler in your hand (Figure 2) and immerse the smaller bulb into a cup of ice water. The large bulb will become very cold as the boiler is set into the ice water.

4. Gently swirl the hand boiler as you warm the large bulb with your cupped hand. Pass the hand boiler and ice-water bath around to let other students swirl the bulb and warm it with their hands.

 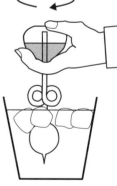

Figure 1.   **Figure 2.**

5. After a few seconds, a colorless liquid will begin to condense in the smaller bulb—simple distillation is taking place.

Teaching with Toys – Page 4

Teacher Notes

6. If some of the dye accidentally gets carried over or "bumped" into the smaller bulb, transfer all the liquid back into the larger bulb and start over.

7. After 10–15 minutes, the larger bulb will contain only the solid, colored dye and the smaller bulb will contain a colorless liquid (see Figures 3 and 4).

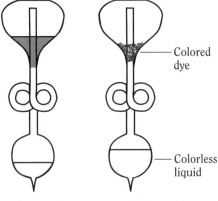

Figure 3.   **Figure 4.**

Disposal

None required—save all materials for future use.

Tip

- Special thanks to Jenna Becker, daughter of Bob Becker, who discovered the phenomenon that made this demonstration possible.

Discussion

The vapor pressure of a liquid is the pressure of the gas in equilibrium with the liquid at a specific temperature. Vapor pressure is a physical property that indicates the tendency of molecules in the liquid phase to evaporate. Evaporation is an endothermic process—energy is required for molecules to overcome the intermolecular attractive forces within the liquid phase and "escape" into the gas phase. Water, which has very strong intermolecular forces, has a very low vapor pressure compared to other liquids with similar molar masses. Hydrocarbons, such as pentane, and other nonpolar liquids have very weak intermolecular forces and correspondingly high vapor pressures (at a given temperature).

The liquid in the hand boiler is a volatile organic solvent—it has weak intermolecular forces and a high vapor pressure. There is also a slight vacuum in the hand boiler. The combination of a liquid with a high vapor pressure in a container that is under a slight vacuum means that the liquid inside the large bulb will evaporate and then boil when only a small amount of heat is added to the system (i.e., the heat from a cupped hand held over the large bulb).

In this demonstration, the liquid in the large bulb begins to boil almost as soon as the small (lower) bulb is placed in an ice–water bath. As the temperature of the smaller bulb is reduced, the pressure of the vapor in the lower bulb also decreases. The opposite effect occurs in the larger (upper) bulb. The added heat from a warm hand increases the vapor pressure of the liquid in the upper bulb. The liquid in the larger bulb is converted to vapor, which condenses in the cooled, smaller bulb. The dye is nonvolatile—it does not evaporate and remains behind in the larger bulb.

The process of vaporizing a liquid in one vessel, condensing the vapor, and then collecting the condensate in a separate vessel is called distillation. Distillation is a commonly used technique in organic chemistry labs and in the chemical process industry (e.g., the petroleum refining industry). It is used to purify compounds and to separate the components in a mixture. See the "Simple Distillation" demonstration in *Elements, Compounds, and Mixtures,* Volume 2 in the *Flinn ChemTopic™ Labs* series, for a discussion of the principles involved in simple distillation and a sample procedure.

Teacher Notes

Demonstrations

Hot Wax
Heat of Fusion of Paraffin

Introduction

The temperature at which paraffin (candle wax) melts is only about 55 °C. When the melted wax solidifies, however, it releases heat and causes severe skin burns. How much heat is released when "hot wax" solidifies?

Concepts

- Phase changes
- Heat of fusion
- Calorimetry

Materials

Paraffin or candle wax, 10 g

Balance, centigram precision (0.01-g)

Beakers, 250- and 400-mL

Boiling stones

Digital thermometers, 2

Graduated cylinder, 250-mL

Hot plate or Bunsen burner setup

Paper towels

Ring stand and clamp

Stirring rod

Styrofoam® cups, 9-oz, 2

Test tube, large, 25 × 150 mm

Safety Precautions

Exercise care when working with hot water baths and hot melted wax. Wear chemical splash goggles, chemical-resistant gloves, and a chemical-resistant apron. Please review current Material Safety Data Sheets for additional safety, handling, and disposal information.

Procedure

1. Fill a 400-mL beaker about two-thirds full with hot tap water. Add a couple of boiling stones to the water and heat to about 90 °C on a hot plate.

2. Add about 100 mL of cold tap water (15–20 °C) to a Styrofoam cup and nest the cup inside a second Styrofoam cup. Place the nested Styrofoam cups in a 250-mL beaker on a ring stand.

3. Obtain about 10 g of paraffin or candle wax shavings in a large test tube and record the precise mass in a data table. Holding the test tube with a clamp or test tube holder, place the paraffin in the hot water bath at about 90 °C and insert a digital thermometer into the paraffin. When the temperature is about 80–85 °C, remove the test tube from the hot water bath and clamp the test tube to the ring stand.

4. Measure and record the precise temperature of the melted paraffin and *immediately* lower the test tube into the cold water bath in the Styrofoam cup. *Start timing.* Carefully stir the paraffin with the digital thermometer and measure the temperature every 30 seconds for 10 minutes, or until the temperature is about 40 °C (whichever comes first). Record all temperature and time measurements in a data table.

5. Plot the cooling curve data for paraffin and estimate its melting point.

Hot Wax

Demonstrations

6. Replace the test tube containing the solidified paraffin back into the hot water bath (step 1).

7. Empty the water from the Styrofoam cup and switch the positions of the nested Styrofoam cups. Obtain about 125 mL of cold tap water (15–20 °C) in a graduated cylinder and measure and record the precise volume in a data table. Carefully pour the water into the Styrofoam cup and place a clean digital thermometer in the cold water.

8. When the paraffin (step 6) has remelted, remove the test tube from the hot water bath and clamp the test tube to the ring stand. Dry the test tube with paper towels.

9. Observe the melted paraffin—when the first traces of solid appear, measure the initial temperature of the cold water and immediately immerse the test tube into the cold water bath. *Note:* The temperature of the paraffin should be very close to its estimated melting point.

10. Carefully stir the paraffin until it solidifies and the temperature of the cold water stabilizes (3–4 minutes). Measure and record the final water temperature and calculate the heat of fusion of paraffin from the calorimetry data.

Disposal

Please consult your current *Flinn Scientific Catalog/Reference Manual* for general guidelines and specific procedures governing the disposal of laboratory wastes. Paraffin may be disposed of according to Flinn Suggested Disposal Method #26a.

Results and Discussion

The melting point of paraffin is 57–58 °C. This is consistent with a straight chain (normal) alkane having the formula $C_{26}H_{54}$ or $C_{27}H_{56}$. Most long-chain *n*-alkanes (> 20 C atoms) exhibit two very closely spaced phase transitions at or just below the melting point. (The second transition corresponds to two crystalline phases, α and β.) Each transition is associated with an enthalpy change. For *n*-$C_{26}H_{54}$, the phase transitions occur at 53 °C and 56.4 °C, and the corresponding enthalpy changes are 21.6 cal/g and 38.6 cal/g, respectively. The literature value for the "heat of fusion" *as determined in this experiment* should be 60.2 cal/g. The experimental value is 61 cal/g (1% error).

Cooling Curve

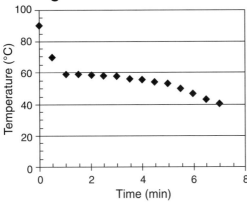

Calorimetry Data

Mass of paraffin	10.08 g
Mass of water	121 g
Water temperature (initial)	17.7 °C
Water temperature (final)	22.8 °C
Temperature change	5.1 °C

$$Q = (121 \text{ g})(5.1 \text{ °C})\left(\frac{1 \text{ cal}}{\text{g} \cdot \text{°C}}\right) = 617 \text{ cal}$$

$$\Delta H_{\text{fusion}} = \frac{617 \text{ cal}}{10.08 \text{ g}} = 61 \text{ cal/g}$$

Teacher Notes

There are two main sources of experimental error in this procedure. (1) The paraffin is placed into the water bath (step 9) too soon. (2) The solid paraffin is allowed to cool in the water bath before the final temperature is recorded (step 10). Both of these errors will lead to a higher calculated heat of fusion.

Demonstrations

Teacher Notes

"Tennis Ball" Distillation
Kinetic-Molecular Theory in Action

Introduction

When molecules are the size of tennis balls, phase changes such as melting and evaporation can be very "moving" events!

Concepts

- Melting and freezing
- Evaporation and condensation

Materials

Balance

Clear plastic tub or container, large enough to hold the tennis balls

Methyl alcohol, 1 mL

Syringe with needle, large

Tennis balls, two different colors, approximately 24 total

Water

Safety Precautions

Make sure that you are aware of your surroundings when shaking the tub. The flying tennis balls could cause damage. Wearing safety glasses or goggles is a good idea.

Preparation

Using a large syringe with needle, inject water (with a couple of drops of methyl alcohol added) into one color of tennis ball. Try to get as much water as possible into the ball. There is no need to seal the ball—the latex in the ball will self-seal and hold the water in. Pretreat about one-half of the tennis balls this way, but don't let the students know this was done.

Procedure

1. Explain to students that each tennis ball represents a molecule in this demonstration.

2. Place all the tennis balls inside the plastic tub. Tilt the tub to show the students that the tennis balls are touching and not moving. *The molecules are in the solid state, locked in fixed positions.*

3. Add some energy to the tennis balls by gently shaking the tub sideways. The tennis balls may still be touching one another, but they will have more motion and become more "fluid." *At first, as the "molecules" gain energy, they begin to rotate and vibrate. When sufficient energy has been added, the molecules will begin to move apart and "break loose" from their rigid positions. This represents the liquid state.*

4. What happens if you stop shaking the tub? Slow down the shaking until all the tennis balls are again stationary. *If energy is not continually added to the tub, the "molecules" will lose energy as they collide and will slow down and "solidify" (freeze) again.*

Demonstrations

Teacher Notes

5. Shake the tub once again, but with even more energy and vigor—the tennis balls will begin to fly out of the tub! (The tub may have to be tilted a little to assist the "evaporation" process. Remember, evaporation requires a lot of energy.) *This represents evaporation. The "molecules" leave the liquid state and enter the vapor phase.*

6. If pretreated tennis balls that are a different color have been used, it should be obvious that only one type of tennis ball has "evaporated." The tennis balls that were injected with water should remain in the tub. *Different compounds may be separated by evaporation followed by condensation of the "more volatile" molecules. This is the process of distillation.*

7. Count on your students' natural curiosity to speculate why only some of the tennis balls have "evaporated." Weigh one of the tennis balls remaining in the tub and compare its mass with one of the tennis balls that "evaporated." *The two sets of "molecules" have different molar masses! The lighter molecules will evaporate, leaving the "heavier" molecules in the liquid state (assuming the molecules are otherwise similar).*

8. Ask students to pick up the tennis balls that have "evaporated" and toss them back into the tub. (What a chance to throw something at the teacher—this may be a good time to put on safety goggles!) Catch the tennis balls in the tub—this represents "condensation." *If the students throw the balls too hard, the "molecules" will have too much energy and will not condense.*

Disposal

None required—save all materials for future use.

Tips

- Different color tennis balls are better than different brands to demonstrate that different substances evaporate at different rates. Talk to the tennis coach at the school about saving old, used tennis balls for the demonstration.

- The number of tennis balls required will depend on the size of the container. Old dish-washing tubs or rectangular plastic bins are good containers for this demonstration. Flinn Scientific sells a sterilizing tray (Catalog No. AP5415) and a large storage container with lid (Catalog No. AP5909) that are suitable for this demonstration.

- Use your imagination to come up with other applications for this truly "kinetic" demonstration of the kinetic-molecular theory. For example, you may want to illustrate that when the tennis balls are in the "solid" state, they exist in a closest-packed arrangement or crystal lattice.

- Please see the *Background* sections in "It's Just a Phase" and "Properties of Liquids" in this lab manual for a good discussion of the kinetic-molecular theory.

- Special thanks to Doug De La Matter, retired chemistry teacher from Madawaska Valley District High School, Barry's Bay, Ontario, for providing Flinn Scientific with the idea for this activity.

Demonstrations

Teacher Notes

Surface Tension Jar

Introduction

Surface tension is a force—a force powerful enough to prevent water from spilling out of an open jar when it is turned upside-down! A fine mesh screen hidden inside the lid of the jar provides hundreds of tiny surface tension "membranes" that, in addition to air pressure, will support the weight of the water.

Concepts

- Surface tension
- Air pressure
- Adhesion vs. cohesion

Materials

Canning jar with screw-on ring lid, 1-qt
Laminated card, 4 in. square
Liquid detergent (optional)
Mesh window screening, 4 in. square
Plastic tub or bucket
Scissors
Tap water

Safety Precautions

Although the materials used in this demonstration are considered nonhazardous, please observe all normal laboratory safety guidelines.

Preparation

Use the removable ring inside the canning lid to trace a circle on the window screen. Cut the screen to size so it will fit inside the screw-on lid. Place the screen between the metal ring and the jar and screw the lid tightly onto the canning jar.

Procedure

1. Pour tap water through the screen until the jar is about three-quarters full.

2. Place a laminated card over the top of the jar and hold the card down tightly with one hand. *The water will form an adhesive seal with the laminated paper.*

3. Quickly invert the jar 180° over a sink or other container, such as a plastic tub or bucket.

4. While holding the jar steady, remove your hand from the laminated card. *The card will remain in place over the mouth of the jar! The water forms a tight adhesive seal and external air pressure holds the card in place.*

5. Carefully slide the card out from under the jar with one hand while holding the jar steady with the other hand. *A little water may spill out, but most of the water will stay in the jar! The mesh screen provides a surface for the formation of hundreds of tiny surface-tension "membranes" that, in addition to air pressure, will support the weight of the water.*

6. Tilt the jar a few degrees to allow air to enter the jar. *The water will immediately spill out of the jar—gravity still works!*

The "Surface Tension Jar" is available as a demonstration kit from Flinn Scientific (Catalog No. AP6648).

Demonstrations

7. *(Optional)* After performing the demonstration once for the students, ask for a student volunteer to repeat the demonstration. Dip a finger into detergent that is hidden from view and inconspicuously run the finger over the screen after the jar has been filled with water. *When the student inverts the jar, the laminated card may stick for a short time due to the counter-force of air pressure acting on the outside of the card. When the card is removed, however, the water will rush out. The detergent interferes with the hydrogen-bonding network in water, which drastically reduces the surface tension of water and modifies its adhesive properties.*

Disposal

None required—save all materials for future use.

Tips

- Window screening material may be obtained from old window screens or purchased at a hardware store.

- One-quart or one-pint home canning jars (also called Mason jars) are terrific for this demonstration. The lid consists of a two-piece vacuum cap containing an outer ring and a removable center disk. Only the outer ring is needed for this demonstration.

Discussion

There are two main questions in this demonstration. What force(s) hold the laminated card in place under the inverted jar (step 4)? What force(s) prevent the water from spilling out when the card is removed (step 5)?

Water is a unique liquid—the surface tension of water is substantially greater than that of alcohols and other liquids with similar properties. The demonstration does not work with alcohols. External air pressure also plays a key "supporting" role. The demonstration will not work if the pressure outside the jar is reduced (for example, if the inverted jar is placed inside a bell jar and a slight vacuum is applied).

When the jar is first inverted, a small amount of water probably leaks out from the jar. This creates a slight partial vacuum in the space above the water in the jar. The water in the jar forms a tight adhesive seal with the card—in addition to forming strong intermolecular "cohesive" forces with other water molecules, water also forms strong "adhesive" forces to many other materials. External air pressure, which acts in all directions, applies a net upward force on the card and water and prevents the water from spilling out of the jar.

When the card is removed, the surface tension of water provides an additional force keeping the water in the jar. The high surface tension of water arises because of strong hydrogen bonding among water molecules. Uneven attractive forces between water molecules at the surface of the liquid versus those in the rest of the liquid result in a net attractive force that tends to "pull" adjacent surface molecules inward to the rest of the liquid. An effective analogy is that surface tension acts as an invisible, elastic film or membrane that "expands" as needed to counteract the force of gravity. The numerous tiny holes in the screen provide a larger total area for the formation of this elastic film.

Teacher Notes

Teacher Notes

Freezing by Boiling
Discrepant Event

Introduction

The boiling point of a liquid depends on the external air pressure. When water is placed under vacuum, the boiling point decreases and the water boils. Boiling, however, is an endothermic process—as the water boils, the temperature decreases, and the water soon freezes!

Concepts

- Boiling point
- Vapor pressure

Materials

Acetone solution, 60% in water, 10 mL

Boiling stones, 2

Construction paper, black

Flinn ChemCam™ video camera (optional)

Vacuum pump with vacuum tubing and 3-way valve

Vacuum plate and bell jar (vacuum chamber)

Pipet, Beral-type

Plastic wrap, 8 × 8 in.

Styrofoam® cup, 8-oz

Safety Precautions

Check the bell jar or vacuum chamber for cracks or chips before use—never place a chipped or cracked jar under vacuum. Placing all items under vacuum behind a safety shield is recommended. Acetone is a flammable liquid and mildly toxic by ingestion and inhalation. Wear chemical splash goggles, chemical-resistant gloves, and a chemical-resistant apron. Please consult current Material Safety Data Sheets for additional safety, handling, and disposal information.

Procedure

1. Cut a piece of black construction paper to line the inside of the Styrofoam cup. (Boiling and freezing will be more visible against this dark background.)

2. Place a piece of plastic wrap over the mouth of the cup and push the plastic wrap down into the cup to create a small "well" as shown in Figure 1. (The well will prevent the liquid from splattering and also make it easier to see the phase changes.)

3. Mix 6 mL of acetone with 4 mL of deionized water to form an aqueous solution.

4. Carefully add about 4–5 mL of the acetone solution into the "well" formed by the plastic wrap and add several boiling stones to prevent "bumping."

5. Set the cup on the vacuum plate (don't cover the hole) and place the bell jar or vacuum chamber over the cup.

Figure 1.

Demonstrations

Teacher Notes

6. Start the vacuum pump and close the valve to evacuate the vacuum chamber.

7. Observe the phase changes for the aqueous solution. *Within seconds, the acetone solution will start to boil. After a few minutes, the solution will start to freeze, but there will be boiling bubbles visible under the ice. Boiling and freezing will occur simultaneously for at least five minutes! Some of the bubbles seem to "explode" into tiny pieces of ice.*

8. Slowly open the three-way valve to release the vacuum in the bell jar, and then turn off the vacuum pump.

Disposal

Please consult your current *Flinn Scientific Catalog/Reference Manual* for general guidelines and specific procedures governing the disposal of laboratory wastes. Allow the acetone solution to evaporate overnight and dispose of the remaining water down the drain.

Tips

- Make the well in the plastic wrap as deep as possible so the solution does not escape. The more liquid in this well, the easier it is to see the boiling and freezing.

- Set up a Flinn ChemCam video camera to give students a close-up view of special effects!

Discussion

Vaporization is the process by which a substance changes from a liquid to a gas or vapor. When vaporization occurs gradually from the surface of a liquid, it is called evaporation. The pressure of the vapor in equilibrium with the liquid at a specific temperature is the vapor pressure. When the vapor pressure equals the external pressure, vaporization can occur throughout the liquid, not just at the surface. Bubbles of vapor then form in the liquid and rise to the surface—the liquid boils. The boiling point of a liquid is defined as the temperature at which the vapor pressure of a liquid is equal to the external (atmospheric) pressure. Thus, the boiling point of a liquid depends on the external pressure. Since the vapor pressure of a liquid always increases as the temperature increases, a liquid will boil at a lower temperature when the external pressure is reduced.

Evaporation is an endothermic process—energy in the form of heat is required for molecules to leave the liquid phase and enter the gas phase. The most common way to provide energy for the vaporization of a liquid is by heating it. When the heat energy for vaporization comes from the surroundings rather than from external heating, however, the temperature of the liquid will decrease as it evaporates. Thus, a liquid cools as it evaporates.

In this demonstration, the acetone solution in the cup begins to boil at room temperature almost as soon as the external pressure is reduced (under vacuum). The temperature of the solution decreases, and when the liquid gets cold enough, it freezes—boiling and freezing occur simultaneously! The acetone–water solution has a higher vapor pressure than pure water and the acetone boils first. Water, however, freezes at a higher temperature than acetone, so the frozen solid is probably water.

Acetone and water form a minimum boiling azeotrope containing 88% acetone and 12% water.

Demonstrations

Teacher Notes

Wet Dry Ice
Triple Point of CO$_2$

Introduction

The solid form of carbon dioxide is called "dry ice" because it does not melt, as ordinary ice does, but rather goes directly from a solid to a gas—it sublimes. The liquid form of dry ice is not stable at "ordinary" pressures and temperatures (atmospheric conditions). If dry ice is allowed to sublime in a closed system, however, the pressure will increase to a point where the liquid form of carbon dioxide can be seen.

Concepts

- Phase changes
- Phase diagrams
- Triple Point
- Sublimation

Materials

Beral-type pipets, wide-stem, 2
Cup, clear plastic, 8- or 10-oz
Dry ice
Flinn ChemCam™ Video Camera (optional)
Gloves, insulated
Hammer, small
Pliers
Scissors

Use a Flinn ChemCam video camera to provide a close-up of the demonstration.

Safety Precautions

Dry ice is extremely cold and may cause frostbite. Handle only with insulated or heavy cloth gloves and never with wet hands. Do not add more than a small amount of dry ice to the pipet. The demonstrator and all observers must wear chemical splash goggles.

Procedure

1. Pulverize the dry ice into small pieces about the size of rice grains or sugar crystals. *Observe that the dry ice does not melt, it sublimes. The resulting "fog" is due to water vapor condensing on the extremely cold CO$_2$ gas that is produced.*

2. Cut off the tapered end of a wide-stem, Beral-type pipet. Scoop about 8–10 pieces of dry ice into the stem of the pipet and tap the dry ice down into the bulb.

3. Add tap water to a clear plastic cup to a depth of about 4–5 cm.

4. Fold the open end of the pipet stem over and clamp it shut with a pair of pliers. (No gas should be able to escape from the pipet.) Immediately lower the pipet bulb into the water in the cup.

5. Observe the phase changes. *The dry ice will sublime (turn to a gas). After about 20–30 seconds, the dry ice will melt and liquid will appear in the pipet bulb. Soon after, the liquid will begin to boil, and the pipet bulb will swell. Three phases will be visible at the same time (solid, liquid, and gas).*

"Demonstrating Phase Changes of CO$_2$" is available as a demonstration kit from Flinn Scientific (Catalog No. AP4505). The kit contains a triple point apparatus with a pressure gauge and a large valve and provides a larger scale version of this demonstration.

Demonstrations

Teacher Notes

6. Release the grip on the pliers to relieve some of the pressure in the pipet. *A loud pop is produced and the CO_2 immediately returns to a solid—the dry ice looks like fluffy snow.*

7. Repeat the demonstration by folding the open end of the pipet stem over again and reclamping the pipet shut with the pliers. *The CO_2 appears to liquefy quicker than it did the first time. Depending on the amount of dry ice used, the process may be repeated 3–4 times.*

8. Your students will undoubtedly ask: "What will happen if you don't release the pressure?" Repeat the demonstration, but don't release the pressure when the liquid ("wet dry ice") begins to boil. *Caution: Everyone should be wearing goggles as explained in the Safety Precautions. The pipet will continue to expand until the gas "explodes" and the bulb ruptures. Water goes everywhere, the cup cracks, and your nerves may be a little more frayed—but the students think it is great fun!*

Disposal

Allow excess dry ice to sublime and discard the used pipets in the trash.

Tips

- This activity is written as a demonstration, but it is safe enough to be used as a student activity. Monitor students so they do not add too much dry ice to their pipets, and strictly enforce the "wear your goggles rule."

- Wide-stem and extra large-bulb pipets (Flinn Catalog No. AP1721 and AP1720, respectively) work best. Use only plastic cups (plastic will absorb the shock and the cup will crack, not shatter). The water in the cup acts as a heat source for sublimation and melting. It also keeps condensation from forming on the outside of the pipet, which would make it difficult to see the contents.

Discussion

From making "fog" to "boiling in water," dry ice is well-known for creating special effects. At atmospheric pressure, carbon dioxide can exist only as a solid or a gas. In order to exist as a liquid, carbon dioxide must be subjected to a pressure of at least 5.1 atmospheres. The *phase diagram* shows how the phase of carbon dioxide depends on pressure and temperature. The boundaries (lines) between the phases in the phase diagram show the values of pressure and temperature when two phases will be in equilibrium. The point at which all three phase boundaries meet is called the *triple point* and signifies the temperature and pressure at which all three phases exist and are in equilibrium.

If a sample of dry ice is sealed in a closed system, the pressure begins to rise as CO_2 gas is produced. The increased pressure, in turn, allows the solid to exist at a higher temperature. Equilibrium continues to exist between the solid and gaseous CO_2 as the temperature and pressure increase. When the triple point is reached, the solid can now sublime or melt. The solid–gas, solid–liquid, and liquid–gas phases are all in equilibrium. As long as solid, liquid, and gaseous CO_2 are in contact with each other, the temperature and pressure will remain at the triple point of 5.1 atm and –56.6 °C.

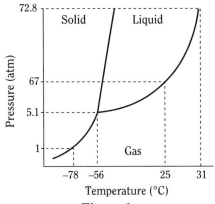

Figure 1.

Demonstrations

Teacher Notes

Four-Square Diffusion
Vapor Pressure of Liquids and Solids

Introduction

A small amount of liquid is placed in one compartment of a partitioned Petri dish, and two solids are sprinkled into adjacent compartments. After a few seconds, one of the solids becomes "wet" and soon dissolves into a puddle of liquid. After a minute or so, the second solid also begins to dissolve into the small puddle of liquid. The rate at which each solid dissolves illustrates how vapor pressure and intermolecular forces influence diffusion rates.

Concepts

- Vapor pressure
- Diffusion
- Sublimation
- Intermolecular attractive forces

Materials

Acetone, 1 mL
Camphene, 5–10 small crystals*
Iodine, 5–10 small crystals
Four-quadrant glass Petri dish with cover
Pasteur pipet or medicine dropper, glass
Spatula or tweezers
Transparency and pens
Overhead projector and screen

Camphor may be used instead of camphene. The demonstration will take a few minutes longer.

Safety Precautions

Camphene and iodine are toxic by ingestion and inhalation—work with these materials in a fume hood or in a well-ventilated laboratory. Camphene and iodine are both irritants to skin, eyes, and respiratory tract. Acetone is a flammable liquid and mildly toxic by ingestion and inhalation. Camphene is a flammable solid; keep away from any sparks or flames. Wear chemical splash goggles, chemical-resistant gloves, and a chemical-resistant apron. Please consult current Material Safety Data Sheets for additional safety, handling, and disposal information.

Procedure

1. Place a four-quadrant glass Petri dish on an overhead projector stage.

2. Use a spatula or tweezers to add 5–10 small crystals of camphene into one compartment, 5–10 small crystals of iodine into the opposite compartment, and one crystal of camphene towards the outer edge of one of the remaining compartments (Figure 1). It is helpful to write the name of each substance and its vapor pressure next to its quadrant on a transparency placed beneath the Petri dish or on the cover.

3. Add about 1 mL of acetone to the last compartment using either a glass pipet or eyedropper.

4. Place the cover on the Petri dish and observe the results.

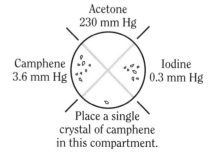

Figure 1.

"Four-Square Diffusion" is available from Flinn Scientific (Catalog No. AP4589). This kit contains the glass partitioned Petri dish and all the chemicals. Special thanks to Walter Rohr, retired teacher from Eastchester High School, Eastchester, NY for providing Flinn Scientific with this idea.

Demonstrations

Teacher Notes

Results

After 15–20 seconds, small droplets of liquid form around the camphene crystals, indicating that the acetone is vaporizing and starting to condense on the camphene crystals. The camphene crystals will quickly dissolve in acetone so that only droplets of solution will be visible in the camphene quadrant after one minute.

After 2–3 minutes, a yellow "halo" will be observed around the edges of the single camphene crystal nearest to the iodine. The yellow color is due to iodine subliming and condensing around the camphere. The camphene–acetone solution will also get a faint yellow tinge, but this will not be as pronounced.

After 3–4 minutes, the iodine crystals will also become wet and dissolve in the acetone. The amount of iodine–acetone solution in this quadrant will continue to increase.

It takes longer for acetone to form a solution with the single camphene crystal at the outer edge of the Petri dish than with the camphene immediately adjacent to the acetone.

Disposal

Please consult your current *Flinn Scientific Catalog/Reference Manual* for general guidelines and specific procedures governing the disposal of laboratory waste. Place the uncovered Petri dish in a fume hood or in well-ventilated area and allow the substances to evaporate. Rinse the Petri dish in the sink with plenty of water.

Tips

- Use only glass Petri dishes—acetone reacts with polystyrene used in plastic Petri dishes.

- Camphene, $C_{10}H_{16}$, is a natural terpene found in pine oil, turpentine, citronella, ginger, nutmeg, etc. Camphor will also work, but the demonstration takes a few minutes longer.

Camphene

Discussion

Intermolecular forces are the weak attractive forces between molecules in the liquid or solid state. A substance that vaporizes easily (has a high vapor pressure) has very weak attractive forces between molecules. A substance with very strong intermolecular forces has strong forces between molecules and thus does not vaporize easily (has a low vapor pressure). Although there can be a wide range in the strength of intermolecular attractive forces between molecules, intermolecular forces are always *significantly* weaker than the covalent bonds between atoms in a molecule.

It is important to recognize that when a compound undergoes a change in state, such as from a solid to a liquid or from a liquid to a gas, only the intermolecular attractive forces (those between molecules) are affected. Covalent bonds linking atoms together in a molecule do not change when a substance undergoes a phase change.

Demonstrations

Teacher Notes

The substances used in this demonstration are covalent compounds with strong covalent bonds linking atoms within the molecules. The attractive forces between the molecules, however, are relatively weak. All three compounds have a relatively high vapor pressure at room temperature—it is easy for the molecules to "break free" from these intermolecular attractive forces and "escape" into the vapor phase.

The nature and strength of the intermolecular forces that exist between molecules depends on the nature of the molecules themselves. Both iodine (I_2) and camphene ($C_{10}H_{16}$) are nonpolar molecules. Iodine has no dipole moment and camphene has only a slight dipole moment. As nonpolar molecules, their intermolecular forces are mainly due to *London dispersion forces*.

Acetone is a polar compound but has a low boiling point due to its low molecular weight. Substances such as acetone are said to be *volatile*—they evaporate rapidly from an open container. These substances have a high *vapor pressure*. Vapor pressure is the pressure of the vapor of a substance above its liquid or solid at equilibrium. If a volatile liquid is placed in a closed container, the liquid will begin to vaporize and the vapor pressure will be the same throughout the closed system.

Acetone, with a vapor pressure of 230 mm Hg, has the highest vapor pressure of the three substances used in this demonstration. Acetone will evaporate readily, saturating the air space with its vapor until the amount of acetone equals a vapor pressure of 230 mm Hg. The iodine and camphene also vaporize (sublime). However, iodine has a vapor pressure of 0.3 mm Hg and camphene has a vapor pressure of 3.6 mm Hg. In other words, fewer molecules of iodine leave the surface of the solid than camphene. The reason for this phenomenon is that iodine has a greater molecular mass (254 g/mole) than camphene (136 g/mole) and shows greater London dispersion forces, thus requiring a greater amount of energy to separate its molecules. Therefore, the vapor pressure of iodine is lower.

The solution is formed when migrating acetone molecules collide with the surface of the camphene. Since the vapor pressure of the camphene is lower than that of pure acetone, the resulting mixture will have a lower vapor pressure than pure acetone (due to Raoult's Law), preventing equilibrium from being achieved. As a result, higher vapor pressure acetone molecules continue to be forced into the lower vapor pressure solution in an attempt to reach equilibrium. A similar scenario also occurs with the iodine crystals but at a slower rate.

If the demonstration is allowed to continue, this process will proceed until most of the acetone has condensed to make camphene and iodine solutions. Eventually, the iodine can be observed discoloring the acetone at the edge nearest its compartment and itself being wet with the acetone.

The purpose of the lone crystal in the bottom cell is to show that being farther away from the acetone, it will take longer for a solution to form. Also, since it is closer to the iodine, it will be more strongly affected by the diffusing iodine vapor.

Safety and Disposal Guidelines

Safety Guidelines

Teachers owe their students a duty of care to protect them from harm and to take reasonable precautions to prevent accidents from occurring. A teacher's duty of care includes the following:

- Supervising students in the classroom.
- Providing adequate instructions for students to perform the tasks required of them.
- Warning students of the possible dangers involved in performing the activity.
- Providing safe facilities and equipment for the performance of the activity.
- Maintaining laboratory equipment in proper working order.

Safety Contract

The first step in creating a safe laboratory environment is to develop a safety contract that describes the rules of the laboratory for your students. Before a student ever sets foot in a laboratory, the safety contract should be reviewed and then signed by the student and a parent or guardian. Please contact Flinn Scientific at 800-452-1261 or visit the Flinn Website at www.flinnsci.com to request a free copy of the Flinn Scientific Safety Contract.

To fulfill your duty of care, observe the following guidelines:

1. **Be prepared.** Practice all experiments and demonstrations beforehand. Never perform a lab activity if you have not tested it, if you do not understand it, or if you do not have the resources to perform it safely.

2. **Set a good example.** The teacher is the most visible and important role model. Wear your safety goggles whenever you are working in the lab, even (or especially) when class is not in session. Students learn from your good example—whether you are preparing reagents, testing a procedure, or performing a demonstration.

3. **Maintain a safe lab environment.** Provide high-quality goggles that offer adequate protection and are comfortable to wear. Make sure there is proper safety equipment in the laboratory and that it is maintained in good working order. Inspect all safety equipment on a regular basis to ensure its readiness.

4. **Start with safety.** Incorporate safety into each laboratory exercise. Begin each lab period with a discussion of the properties of the chemicals or procedures used in the experiment and any special precautions—including goggle use—that must be observed. Pre-lab assignments are an ideal mechanism to ensure that students are prepared for lab and understand the safety precautions. Record all safety instruction in your lesson plan.

5. **Proper instruction.** Demonstrate new or unusual laboratory procedures before every activity. Instruct students on the safe way to handle chemicals, glassware, and equipment.

Safety and Disposal

6. **Supervision.** Never leave students unattended—always provide adequate supervision. Work with school administrators to make sure that class size does not exceed the capacity of the room or your ability to maintain a safe lab environment. Be prepared and alert to what students are doing so that you can prevent accidents before they happen.

7. **Understand your resources.** Know yourself, your students, and your resources. Use discretion in choosing experiments and demonstrations that match your background and fit within the knowledge and skill level of your students and the resources of your classroom. You are the best judge of what will work or not. Do not perform any activities that you feel are unsafe, that you are uncomfortable performing, or that you do not have the proper equipment for.

Safety Precautions

Specific safety precautions have been written for every experiment and demonstration in this book. The safety information describes the hazardous nature of each chemical and the specific precautions that must be followed to avoid exposure or accidents. The safety section also alerts you to potential dangers in the procedure or techniques. Regardless of what lab program you use, it is important to maintain a library of current Material Safety Data Sheets for all chemicals in your inventory. Please consult current MSDS for additional safety, handling, and disposal information.

Disposal Procedures

The disposal procedures included in this book are based on the Suggested Laboratory Chemical Disposal Procedures found in the *Flinn Scientific Catalog/Reference Manual*. The disposal procedures are only suggestions—do not use these procedures without first consulting with your local government regulatory officials.

Many of the experiments and demonstrations produce small volumes of aqueous solutions that can be flushed down the drain with excess water. Do not use this procedure if your drains empty into groundwater through a septic system or into a storm sewer. Local regulations may be more strict on drain disposal than the practices suggested in this book and in the *Flinn Scientific Catalog/Reference Manual*. You must determine what types of disposal procedures are permitted in your area—contact your local authorities.

Any suggested disposal method that includes "discard in the trash" requires your active attention and involvement. Make sure that the material is no longer reactive, is placed in a suitable container (plastic bag or bottle), and is in accordance with local landfill regulations. Please do not inadvertently perform any extra "demonstrations" due to unpredictable chemical reactions occurring in your trash can. Think before you throw!

Finally, please read all the narratives before you attempt any Suggested Laboratory Chemical Disposal Procedure found in your current *Flinn Scientific Catalog/Reference Manual*.

Flinn Scientific is your most trusted and reliable source of reference, safety, and disposal information for all chemicals used in the high school science lab. To request a complimentary copy of the most recent *Flinn Scientific Catalog/Reference Manual,* call us at 800-452-1261 or visit our Web site at www.flinnsci.com.

National Science Education Standards

Experiments and Demonstrations

Content Standards	It's Just a Phase	Properties of Liquids	Vapor Pressure of Water	How Cool Is That?	Teaching with Toys	Hot Wax	"Tennis Ball" Distillation	Surface Tension Jar	Freezing By Boiling	Wet Dry Ice	Four-Square Diffusion
Unifying Concepts and Processes											
Systems, order, and organization	✓	✓	✓	✓			✓				
Evidence, models, and explanation	✓	✓	✓	✓	✓	✓	✓	✓	✓	✓	✓
Constancy, change, and measurement	✓	✓	✓	✓							
Evolution and equilibrium									✓		✓
Form and function											
Science as Inquiry											
Identify questions and concepts that guide scientific investigation	✓	✓	✓	✓	✓	✓		✓	✓	✓	✓
Design and conduct scientific investigations	✓	✓	✓	✓	✓	✓					
Use technology and mathematics to improve scientific investigations	✓			✓							
Formulate and revise scientific explanations and models using logic and evidence	✓			✓	✓			✓	✓	✓	✓
Recognize and analyze alternative explanations and models											
Communicate and defend a scientific argument											
Understand scientific inquiry											
Physical Science											
Structure of atoms											
Structure and properties of matter	✓	✓	✓	✓	✓	✓	✓	✓	✓	✓	✓
Chemical reactions											
Motions and forces								✓			
Conservation of energy and the increase in disorder											
Interactions of energy and matter											

National Science Education Standards

Experiments and Demonstrations

Content Standards (continued)	It's Just a Phase	Properties of Liquids	Vapor Pressure of Water	How Cool Is That?	Teaching with Toys	Hot Wax	"Tennis Ball" Distillation	Surface Tension Jar	Freezing By Boiling	Wet Dry Ice	Four-Square Diffusion
Science and Technology											
Identify a problem or design an opportunity											
Propose designs and choose between alternative solutions											
Implement a proposed solution											
Evaluate the solution and its consequences											
Communicate the problem, process, and solution											
Understand science and technology											
Science in Personal and Social Perspectives											
Personal and community health			✓								
Population growth											
Natural resources											
Environmental quality											
Natural and human-induced hazards											
Science and technology in local, national, and global challenges											
History and Nature of Science											
Science as a human endeavor											
Nature of scientific knowledge											
Historical perspectives											

Master Materials Guide

(for a class of 30 students working in pairs)

Experiments and Demonstrations

	Flinn Scientific Catalog No.	It's Just a Phase	Properties of Liquids	Vapor Pressure of Water	How Cool Is That?	Teaching with Toys	Hot Wax	"Tennis Ball" Distillation	Surface Tension Jar	Freezing By Boiling	Wet Dry Ice	Four-Square Diffusion
Chemicals												
Acetone	A0009			25 mL						6 mL		1 mL
Boiling chips	B0136						2			2		
Camphor	C0354											1 g
Cleaner, dishwashing	C0241								1			
Dodecyl sulfate, sodium salt	D0040		1 g									
Ethyl alcohol, anhydrous	E0012			25 mL								
n-Heptane	H0051			25 mL								
Hexanes	H0046			25 mL								
Iodine	I0006											1 g
Isopropyl alcohol	I0019		15 mL	25 mL								
Isopropyl alcohol, 70%	I0021		21 mL									
Lauric acid	L0052	90 g										
Methyl alcohol	M0054			25 mL					1 mL			
Paraffin wax	P0003						10 g					
Glassware												
Beakers												
250-mL	GP1020	15				2	1					
400-mL	GP1025	15					1					
1-L, tall-form	GP1061			15								
Graduated cylinders												
25-mL	GP2010			15								
100-mL	GP2020	15										
250-mL	GP2025								1			
Ointment jar, 120-mL	AP8445		7									
Pasteur pipet, glass	GP7042											1
Stirring rod	GP5075								1			
Test tubes												
13 × 100 mm	GP6063				60							
18 × 150 mm	GP6067	15										
25 × 150 mm	GP6069								1			

Continued on next page

Master Materials Guide

(for a class of 30 students working in pairs) — **Experiments and Demonstrations**

General Equipment and Miscellaneous

Item	Flinn Scientific Catalog No.	It's Just a Phase	Properties of Liquids	Vapor Pressure of Water	How Cool Is That?	Teaching with Toys	Hot Wax	"Tennis Ball" Distillation	Surface Tension Jar	Freezing By Boiling	Wet Dry Ice	Four-Square Diffusion
Balance, centigram (0.01-g precision)	OB2059	3					1	1				
Barometer	AP5070			optional								
Bell jar, large	AP1870					1						
Bell jar, small	AP6655									1		
Capillary tubes, open-ended	GP7046		90									
ChemCam™ video camera	AP4560									optional	optional	optional
Clamp, buret	AP1034	15					1					
Corks, size 2	AP8302				60							
Cups, clear plastic	AP6543										1	
Cups, Styrofoam®	AP1190	30					2			1		
Drinking bird	AP9292					1						
Gloves, autoclave	AP8876			7								
Hammer	AP4436										1	
Hand boiler	AP9293					1						
Hot plate	AP4674	7		7			1					
Filter paper, 11.0 cm diameter	AP8997				60							
LabPro™ Interface System	TC1500				10							
Logger Pro™ software	TC1421				1							
Metric ruler	AP4684		15									
Pipets, Beral-type, graduated	AP1721		90		60							
Pipets, Beral-type, jumbo	AP8850			15								
Pipets, Beral-type, wide-stem	AP2253										1	
Plastic wrap	AP1736									8 in		
Pliers, long-nose	AP8389										1	
Reaction strips, 12-well	AP1446		15									
Ring support with clamp	AP8232			optional								
Rubber bands, orthodontic	AP2008				60							
Scissors, student	AP5394				10					1		
Spatula	AP8338	15										1
Stirring rods, plastic	AP8150			15								
Stoppers, size 2, one-hole	AP2302			15								
Storage container with lid	AP5909							1	1			

Continued on next page

Master Materials Guide

(for a class of 30 students working in pairs)

Experiments and Demonstrations

	Flinn Scientific Catalog No.	It's Just a Phase	Properties of Liquids	Vapor Pressure of Water	How Cool Is That?	Teaching with Toys	Hot Wax	"Tennis Ball" Distillation	Surface Tension Jar	Freezing By Boiling	Wet Dry Ice	Four-Square Diffusion
General Equipment and Miscellaneous, continued												
Support stand	AP8226	15		optional			1					
Syringe with needle, 10-mL	AP1149								1			
Temperature probe	TC1502				20							
Test tube clamp	AP8217	15					1					
Test tube rack	AP1319				10							
Thermometer, digital	AP8716	22		15			2					
Vacuum plate	AP6683									1		
Vacuum pump	AP1597									1		
Vacuum tubing	AP8789									1		
Valve, three-way	AP5353									1		
Water, distilled or deionized	W0007 / W0001		✓	✓						✓		
Weighing dishes	AP1278	15										
Wire gauze	AP1699			optional								

Flinn ChemTopic™ Labs — Solids and Liquids